THE
DINOSAUR
CRAFT
BOOK

THE DINOSAUR CRAFT BOOK

First published in the United Kingdom
by Charles Letts & Co Ltd 1992
Letts of London House
Parkgate Road
London SW11 4NQ

© Complete Editions 1992

Editor: Sandy Ransford
Cover and book design: Craig Dodd
Illustrations: Paul Harrison, David Mostyn, Francesca O'Brien
Knitting grids: Clive Sutherland
Photography: Julie Fisher
Stylist: Simon Lycett

All rights reserved. No part of this publication may be reproduced,
stored in a retrieval system, or transmitted in any form
or by any means, electronic, mechanical, photocopying, recording or
otherwise, without prior written permission of the Publisher.

A CIP catalogue record for this book
is available from the British Library
ISBN 1 85238 335 6

'Letts' is a registered trademark of Charles Letts & Co Ltd

Typeset by Saxon Printing, Derby, UK
Printed in Singapore

Contents

Contributors *6*
Introduction *7*

Dinosaurs to Wear

Woolly Mammoth *10*
In the Swim *14*
Brach-pack *19*
Jewellery from Giants *24*
On the Button *28*
Primeval Painting: Silk *31*

Dinosaurs to Play With

Tyrannosaurus Tex *36*
Katie Plateosaurus *39*
Bethia Polacanthus *42*
Monster Mobile *45*
Pterodactyl Gliders *50*

Dinosaurs in the Home

Bathtime Behemoth *54*
Bookworms *55*
Easy Option *60*
Dimetrodon Draught-stop *63*
Eggheads *66*
Cosy Dimetrodon *72*
Primeval Painting: China *75*
Memosaurus *78*
Stencil-osaurus *80*
Alphabetical *83*

Bedtime Dinosaurs

Hot-water Bird *86*
Sleeping Partner *88*
Bronto-snaurus *92*

Party-time Dinosaurs

Bertie Baryonynx *98*
Monster Mask *100*
Party-time Stegosaurus *105*
Eggs in the Nest *109*

Acknowledgements *111*

Contributors

The items featured in this book were designed and made by the following people:

Juliet Bawden:
Primeval Painting: Silk, page 31; Monster Mobile, page 45; Bathtime Behemoth, page 54; Cosy Dimetrodon, page 72; Primeval Painting: China, page 75.

Vivienne Hall:
Brach-pack, page 19; Bronto-snaurus, page 92.

David Hawcock:
Pterodactyl Gliders, page 50; Bertie Baryonynx, page 98; Monster Mask, page 100.

Philippa Lewis, Emma Lewis, Sophie Lewis:
Eggheads, page 66; Hot-water Bird, page 86; Sleeping Partner, page 88.

Pat Lock:
Party-time Stegosaurus, page 105; Eggs in the Nest, page 109.

Joyce Luckin:
Katie Plateosaurus, page 39; Bethia Polacanthus, page 42.

David Mostyn:
Stencil-osaurus, page 80; Alphabetical, page 83.

Linda O'Brien, Francesca O'Brien, Gladys Ketteringham:
Woolly Mammoth, page 10; In the Swim, page 14; Easy Option, page 60; Dimetrodon Draught-stop, page 63.

Sergio Ransford:
Tyrannosaurus Tex, page 36; Bookworms, page 55; Memosaurus, page 78.

Sarah Spooner:
Jewellery from Giants, page 24; On the Button, page 28.

Introduction

The word 'dinosaur', which means 'terrible lizard', was coined in 1841 by the geologist Richard Owen. Dinosaurs lived between 225,000,000 and 65,000,000 years ago, and no one knows for certain why they disappeared. One thing is sure, though, that since their existence became more widely known, they have fascinated millions of people, and especially children.

This book is a celebration of dinosaurs in craft, with things to make for children, family, friends and the home, using a wide variety of skills and materials. A number of them, such as the jumpers, the toys, the mask, the painted plate, stencils and food, could happily be assembled to form the basis of a dinosaur party, which would be a great hit with the younger members of the family.

All the items are simple to make, and detailed instructions are given. None requires any specialised knowledge, or equipment that isn't easily obtainable or to be found in the average home. Children will enjoy making some of the things, especially with a little adult help.

Wherever possible, diagrams have been reproduced in the actual size of the items, but sometimes they have had to be reduced in size in order to fit them on the page. Where this happens, the scale is given with the diagram. In order to create your own, full-scale design from which the item may be made, draw a grid with the size of squares indicated by the scale (e.g. if 1cm of the grid equals 3cm, then draw a grid with 3cm squares) and transfer the design very carefully square by square from the book on to the grid paper.

The knitting projects which feature motifs each have a grid of small squares in which one square equals one stitch. They are reproduced in full colour, so it is easy to see at a glance which stitch you have arrived at and what colour it should be. Happy dinosaur making!

DINOSAURS TO WEAR

Woolly Mammoth

This jolly jumper, which features a triceratops, has been designed to fit an eleven- to thirteen-year-old.

Materials

- 2 (2-2) 50g balls of double knitting yarn in 1st colour, **green** (A)
- 1 (1-1) ball in 2nd colour, **yellow** (B)
- 5 (5-6) balls in 3rd colour, **turquoise** (C)
- 1 (1-1) ball in 4th colour, **pink** (D)
- A small amount each of 5th colour, **white** (E), and **black**
- 2 black beads for eyes
- 1 pair each of 3¼mm (no. 10) and 4mm (no. 8) knitting needles
- 2 stitch-holders

The quantities of yarn given are based on average requirements and are therefore approximate.

Measurements

To fit bust: 76 (81:86) cm (30 (32:34) in).
Actual measurement: 91 (96:101) cm (36 (38:40) in).
Length: 56 (61:66) cm (22 (24:26) in).
Sleeve length (under arm): 40 (42:44) cm (15½ (16½:17½) in).
Sleeve length (over arm): 56 (61:66) cm (22 (24:26) in).
Figures in brackets refer to the larger sizes. Where only one figure is given this refers to all sizes.

Tension

22 sts and 28 rows to 10cm (4in) on 4mm needles over st.st.

Abbreviations

K = knit; P = purl; st(s) = stitch(es); st.st. = stocking stitch; beg. = beginning; foll. = following; inc. = increase; dec. = decrease; cont. = continue; RS = right side; WS = wrong side; rep. = repeat; rem. = remaining; cm = centimetre; in = inch; tog. = together; SKP = slip one stitch, knit one stitch and pass slipped stitch over.

Method

Back

With 3¼mm needles and A, cast on 68 (74:80) sts and work in K1, P1, rib for 20 rows.
Change to 4mm needles.
Increase row: K2 (5:8), inc. in next st *K1, inc. in next st, rep. from * to last 3 (6:9) sts, K to end of row. (100 (106:112) sts).
Next row: P**.
Now work straight in st.st. for 12 (14:16) rows.

Place chart

Now starting with the 1st row, work from **back** chart, using separate balls of yarn for each colour block, winding yarns around each other at colour changes on every row to stop a hole forming. Work from appropriate lines for size required.
Cont. straight as set until 38th row of chart has been worked, thus ending with a WS row. Work straight in st.st. until back measures 33 (36:38) cm ending with a WS row.

Shape armhole

Cast off 4 (5:6) sts at beg. of next 2 rows.
Dec. 1 st at each end of next and foll. alt. rows until there are 32 sts ending with WS row. Leave sts on holder.

Front

Work as for **back** to ** then work in st.st. for 4 (8:12) rows.

Place chart

Now starting with the 1st row, work from **front** chart, using separate balls of yarn for each colour block, winding yarns around each other at colour changes on every row to stop a hole forming. Work between appropriate lines for size required.
Cont. straight as set until front measures 33 (36:38) cm ending with a WS row.

Shape armhole

Cont. to work from chart but at the same time work shaping as folls.
Cast off 4 (5:6) sts at beg. of next 2 rows.
Dec. 1 st at each end of next and foll. alt. rows until there are 50 sts ending with WS row.

Shape sleeve

Next row: SKP, K 16 put rest of sts onto holder, turn and P to end. Working on these sts only dec. 1 st at each end of next and foll. alt. rows until there is 1 st rem.
Fasten off.
Rejoin yarn to rem. sts and K to last 2 sts, K2tog.
P16, put rem. 14 sts onto holder, turn and work shaping as for other side of neck.

WOOLLY MAMMOTH

Front chart

X = sew beads   - - - - = embroider when work complete

Sleeve

With 3¼mm needles and C, cast on 34 (34:36) sts and work in K1, P1 for 20 rows.
Change to 4mm needles.
Increase row: K4 (1:1), inc. in next 2 sts., *K1, inc. in next 2 sts., rep. from * to last 4 (1:0) sts. K to end. (52 (56:60) sts).
Now starting with a P row work in st. st., but at the same time, inc. 1 st at each end of every foll. 4th row until there are 92 (100:108) sts on the needle.
Now work straight in st. st. until sleeve measures 40 (42:44) cm from cast-on edge, ending with a WS row.

Shape sleeve

Cast off 4 (5:6) sts at beg. of next 2 rows.
Dec. 1 st at each end of next and foll. alt. rows until there are 78 sts ending with WS row.
Dec. 1 st loosely at each end of every row until there are 2 sts.
Cast off.
Join sleeves to front and back, leaving left back seam open.

Neckband

With 3¼mm needles and C, pick up and K 1 st from left sleeve, K 17 sts down left front neck, K 14 sts from front holder, K 17 sts from right front neck, 1 st from right sleeve and finally K 32 sts from back holder (82 sts).
Work in single rib for 9 rows, then starting with a K row work in st. st. for 5 cm.
Cast off loosely.

To make up

Press according to ball-band instructions.
Work outline of legs in B as shown on charts, then with black yarn embroider mouth, and sew beads on for eyes.
Join left back seam and neckband.
Join side and sleeve seams.

Back chart

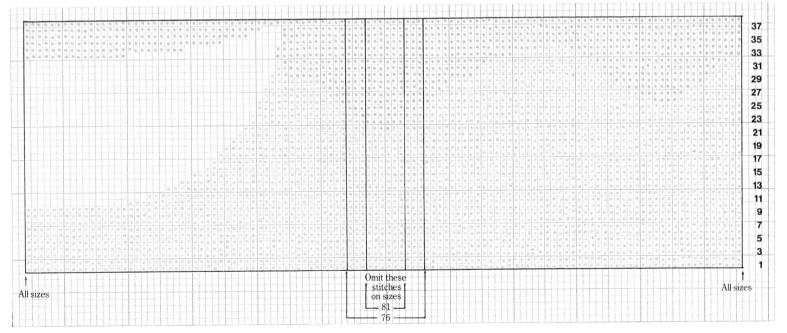

In the Swim

IN THE SWIM

This long-line jumper in adult sizes, which can be worn loose or belted at the waist, features the aquatic dinosaur, plesiosaurus.

Materials

1 (2-2) 50g balls of double knitting yarn in 1st colour, **black** (A)
2 (2-2) balls in 2nd colour, **orange** (B)
2 (2-2) balls in 3rd colour, **royal** (C)
1 (1-1) ball in 4th colour, **yellow** (D)
5 (6-6) balls in 5th colour, **turquoise** (E)
1 (1-1) ball in 6th colour, **dark green** (F)
1 (1-1) ball in 7th colour, **light green** (G)
2 square black beads or buttons for dinosaur eyes
3 small round beads for fish eyes
1 pair each of 3¼mm (no. 10) and 4mm (no. 8) knitting needles
2 stitch-holders

The quantities of yarn given are based on average requirements and are therefore approximate.

Measurements

To fit bust: 86 (91:96) cm (34 (36:38) in).
Actual measurement: 102 (112:122) cm (40 (44:48) in).
Length: 79cm (31in).
Sleeve seam length: 48cm (19in).
Figures in brackets refer to the larger sizes. Where only one figure is given this refers to all sizes.

Tension

22 sts and 28 rows to 10cm (4in) on 4mm needles over st.st.

Abbreviations

K = knit; P = purl; st(s) = stitch(es); st.st. = stocking stitch; beg. = beginning; foll. = following; inc. = increase; dec. = decrease; cont. = continue; RS = right side; WS = wrong side; rep. = repeat; rem. = remaining; cm = centimetre; in = inch; patt. = pattern.

Method

Back
With 4mm needles and A, cast on 112 (122:132) sts and starting with a K row, work in st.st. for 9 rows, ending with RS row. K next row, (this forms the hem turning row).

Place chart
Now starting with the 1st row, work from **back** chart, using separate balls of yarn for each colour block, winding yarns around each other at colour changes on every row to stop a hole forming. Work between appropriate lines for size required.**
Cont. straight as set until 124 rows of chart have been worked, ending with a WS row.
Cont. in st.st. for 8 rows.

Shape armhole
Dec. 1 st at each end of next and foll. 4 alt. rows, (102 (112:122) sts) then cont. straight until work measures 28cm (11in) from beg. of armhole shaping, ending with WS row.

Shape shoulder
Cast off 7 (8:9) sts at beg. of next 6 rows.
Cast off 6 (7:8) sts at beg. of next 4 rows. Leave rem. 36 sts on holder.

Front
Using **front** chart work as for back to **.
Cont. straight as set until 132 rows of chart have been worked, thus ending with a WS row.

Shape armhole
Cont. working from chart but at the same time shape armholes as for back.
When chart has been completed cont. in st.st. until work measures 23cm (9in) from beg. of armhole shaping.

Shape front neck
Next row: K44 (49:54) sts, turn and cont. on this first set of sts only, placing rem. sts on a stitch-holder.
*** Dec. 1 st at neck edge on every row until 33 (38:43) sts remain.

Shape shoulder
Cast off 7 (8:9) sts at beg. of next 3 alt. rows.
Cast off 6 (7:8) sts at beg. of next 2 alt. rows.
With RS facing rejoin yarn to rem. sts and K14. Put these sts onto holder and K to end of row.
Now work as for first side from *** to end.

Sleeves
With 4mm needles and E, cast on 40 sts and starting with a K row, work in st.st. for 9 rows, ending with RS row. K next row, (this forms the hem turning row), then starting with a K row work in st.st. for 10 rows.
Cont. in st.st. but inc. 1 st at each end of next and foll. 3rd rows until there are 96 sts ending with WS row.
Now inc. 1 st at each end of 3rd and foll. 4th rows until there are 116 sts. ending with WS row.

IN THE SWIM

Front chart

X = sew beads

– – – – = embroider when work complete

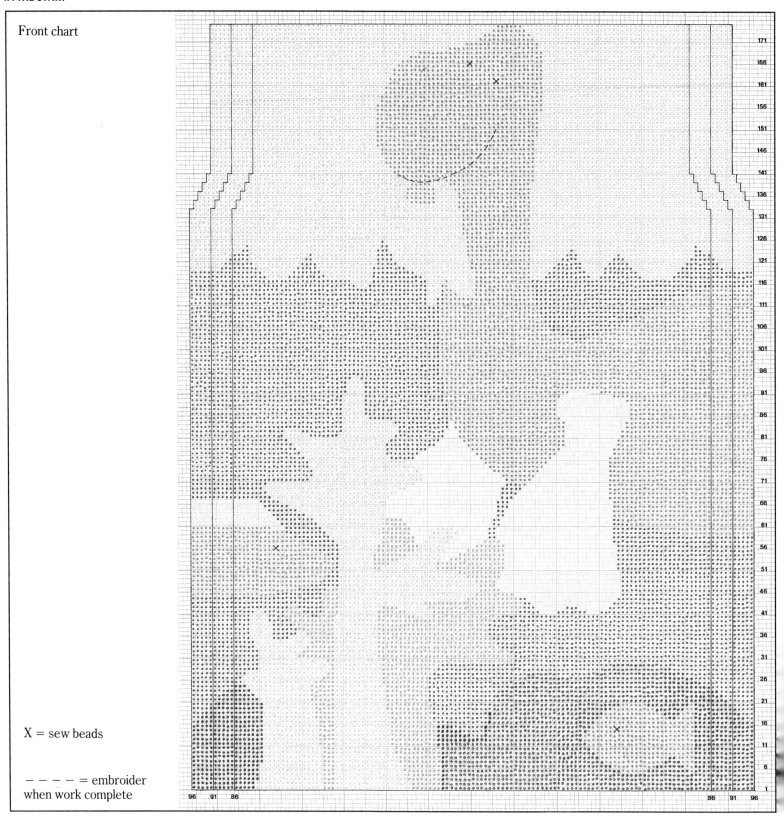

IN THE SWIM

Back chart

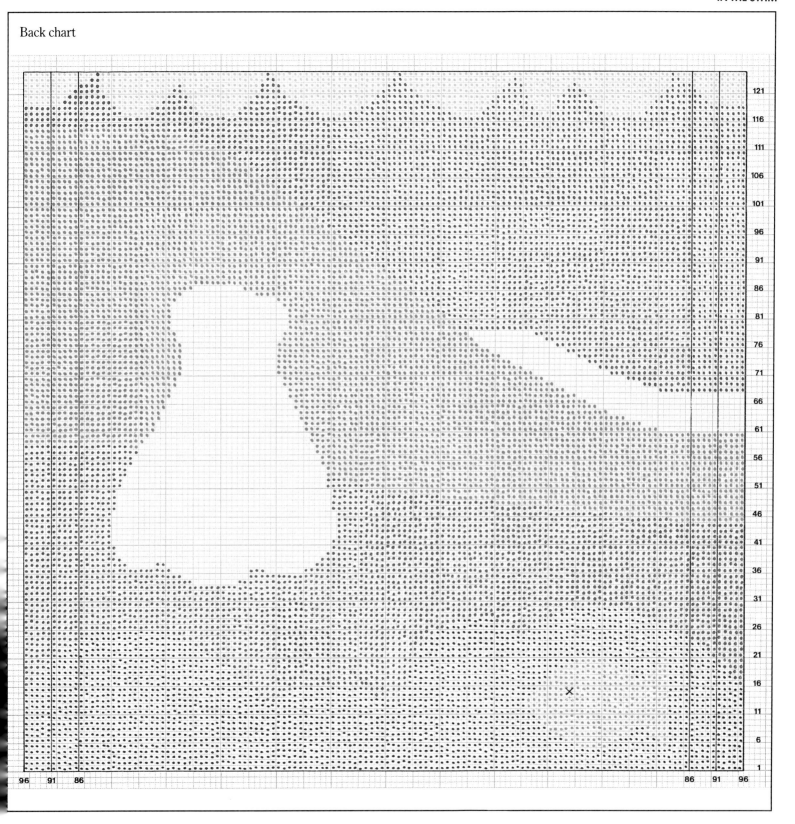

Shape sleeve

Cast off 5 sts at beg. of next 10 rows, then cast off 4 sts at beg. of next 8 rows. Cast off rem. 34 sts.

Neckband

Join right shoulder.
With 3¼mm needles and E and RS facing, pick up and K 18 sts down left front neck, K 14 sts from front stitch-holder, pick up and K 18 sts up right front neck, K 36 sts from back stitch-holder (86sts). Work in K1, P1 rib for 20 rows.
Cast off fairly loosely ribwise.

To make up

Press according to ball-band instructions.
Join left shoulder and neckband seam.
Fold neckband in half to inside and slip-stitch loosely in position.
Sew sleeves into armholes.
Lay jumper out flat with RS up, and using C make raindrops with a long running stitch. Fasten off at beg. and end of each line.
With A embroider mouth, then sew on eyes.
Join side and sleeve seams.
Finally turn up hems on sleeves and body.

Brach-pack

BRACH-PACK

A friendly brachiosaurus is appliquéd onto this colourful padded rucksack to make a delightful present for a young child.

Materials

1m of 120cm wide (1yd of 48in) turquoise polycotton
1m of 120cm wide (1yd of 48in) striped polycotton
0.25m of 90cm wide (¼yd of 36in) dark blue polycotton
1m of 70g (1yd of 2½oz) polyester wadding
2.4m of 2cm wide (2½yd of 1in) dark blue bias binding
2 x 25mm (1in) dark blue toggles
Scraps of pink and yellow polycotton for appliqué
Scrap of interfacing for appliqué
Tailors' chalk

Method

1. Scale up the pattern.
2. Cut out one main section in the turquoise polycotton; one main section in the striped polycotton; one main section in the polyester wadding; two side panels in the turquoise polycotton; two side panels in the striped polycotton; two side panels in the polyester wadding; one strap in the dark blue polycotton; one strap in the polyester wadding.
3. Iron the materials to be used for appliqué onto the interfacing, then draw the shapes on the interfacing. You need one whole body shape in pink (A), and one each of shapes (B) and (C) in yellow.
4. Place the main fabric on the wadding and pin in place, stretching the wadding if necessary. Zig-zag round the edge. Trim any excess wadding.
5. Pin the appliqué shapes onto what will be the flap of the bag, on the right side of the turquoise fabric main section. First pin the pink body in position, and appliqué round the front of the dinosaur from the nose to the tip of the tail using a satin stitch on width 3.
6. Position the yellow shape as shown, pin and stitch, covering the pink edges, which should not be seen.
7. Finally sew the eye with pink thread.
8. To make the strap, place the wadding on top of the dark blue polycotton. Fold the fabric, approximately one-third of its width, over the wadding towards the centre of the strap and pin and tack in place. Then fold the other side over the first fold, turning in the raw edge (about ½cm – ¼in). Tack into position, then straight machine stitch along approximately one-third of the way across the strap. Machine along the other side of the strap to match, thus dividing the strap into three equal sections. Finish off the ends by stitching across several times and trim close to the stitching.
9. To make the side panels, sandwich the side panel wadding between the wrong sides of the turquoise fabric and the striped fabric. Pin securely, then zig-zag round the edges and trim. Repeat for the other panel.
10. To position the strap, measure 27cm (10½in) from the edge of the appliquéd flap, and 13cm (5in) in from the sides and mark with tailors' chalk. Fold the strap in half to form a V shape, and pin to the centre mark of the haversack with the point of the V at the mark.
11. Measure 16cm (6½in) down from the point of the V and mark the haversack. Along this line measure 3cm (1¼in) in from both sides of the haversack and mark these points.
12. Bring both ends of the strap over to these points. Fold the strap edges under for 5cm (2in) and pin with the folded edge underneath.
13. Stitch all three ends of the strap, both round the edges and diagonally, to give it extra strength.
14. To attach the side panels, pin them to the edges of the haversack with the lining fabrics together, taking in the edge of the haversack as you go. Zig-zag in position. Repeat on the other side, then trim.
15. To make the toggle loops, take a piece of dark blue bias binding about 16cm (6in) long and fold it in three, folding inwards to make a long narrow strip. Machine stitch, then cut it in half. Fold each piece in half and pin into place on the inside of the bag flap, with the raw edges facing and quite near to the edge of the flap, so they will be caught in place when the edge is bound.
16. To bind the haversack, open out one folded side of the binding and, starting at the left side panel, machine stitch along the fold line, matching the raw edges (and taking care to conceal any stitches from the making-up). Take care not to stretch the binding as you go round the bottom curves. Take the binding right round the front flap and the second side panel.
17. Fold the binding over to cover the stitch line. Ease the binding round the bottom curves of the side panels (you may find it easier to pin the binding in place here). Stitch the binding over the front all the way round.
18. Next bind the top edge of the bag. Start at the inside of the right-hand panel. Open out one folded edge, and fold back 2cm (1in) at the end to hide the raw edge. Machine stitch round the top edge to the other side, turning back 2cm (1in) at that end, too.
19. Fold the binding to the inside and top stitch, covering the previous stitch line. Neaten the sides by hand, attaching this binding to the main binding.
20. To finish the toggle loops, fold them over towards the front and machine in place with several rows of stitching on the binding.

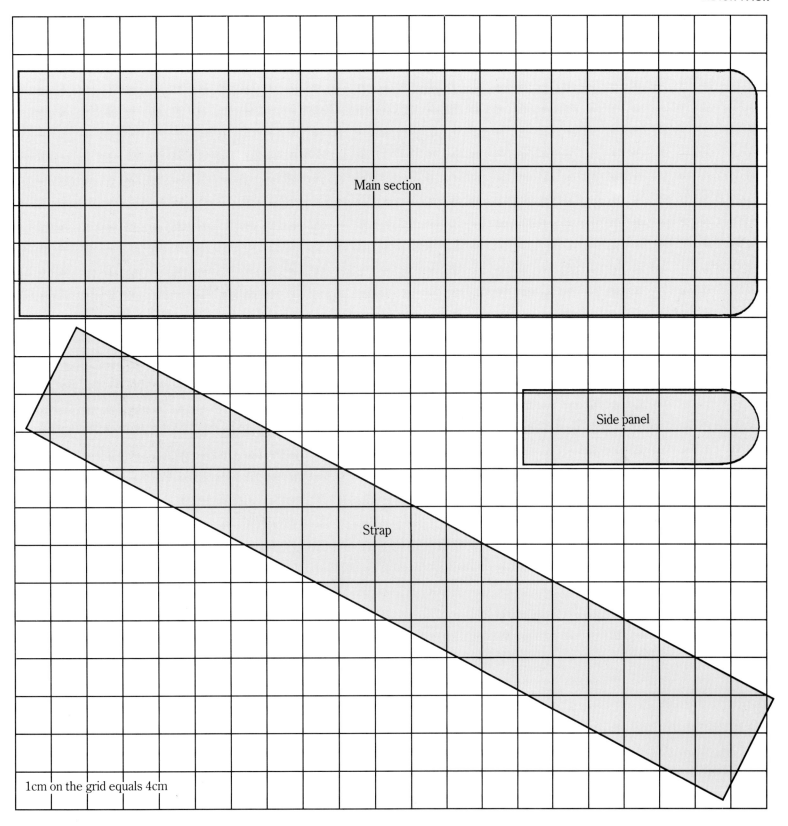

BRACH-PACK

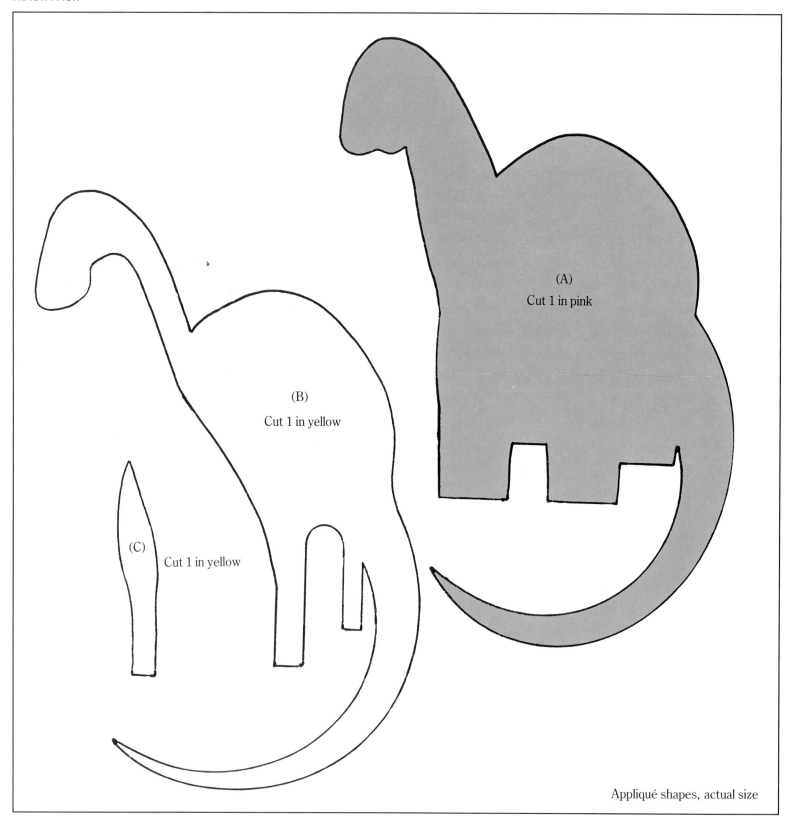

Appliqué shapes, actual size

BRACH-PACK

Strap – stitch across ends to finish off

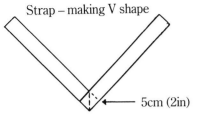
Strap – making V shape
5cm (2in)

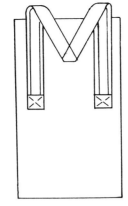
Strap – position of other ends

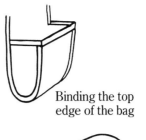
Binding the top edge of the bag

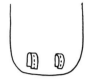
Making toggle loops

Attaching the side panels

Start here

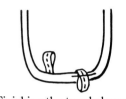
Position of toggle loops

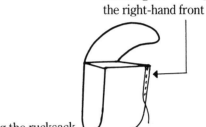
Stitching the binding on the right-hand front

Binding the rucksack

Finishing the toggle loops

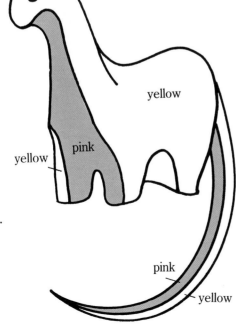
yellow
pink
yellow
pink
yellow

Pink body stitched all the way from nose to tail on front. Yellow pieces placed over pink body. Tacking will ensure no pink is visible round the edges.

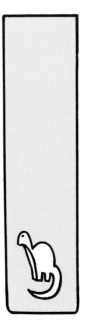

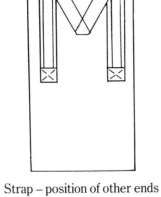
Positioning the appliqué on the front flap.

23

Jewellery from Giants

JEWELLERY FROM GIANTS

The matching earrings, bracelet and necklace are based on the shape of the foot bones of the plesiosaurus, an aquatic dinosaur. They are made from fimo, an easy-to-use craft material, which can be modelled like plasticine and is then baked in the oven to harden it.

Materials

Fimo in turquoise, purple, navy and emerald
Embroidery silk
Modelling knife
Sheet of acetate film or thin, smooth card
Florist's wire
Nylon line
A pencil
Rolling-pin
Cooking oil
Varnish

For earrings:
   2 large jump rings
   2 medium jump rings
   2 pieces kidney wire

For bracelet:
   clasp
   4 medium jump rings

For necklace:
   clasp
   4 medium jump rings
   20-40 glass beads

Method

Earrings

1. Trace off the designs and make acetate or card stencils of them. Acetate is better because it is easier to peel off the fimo, and also because you can see through it. Wipe it lightly with a cloth greased with cooking oil before putting the stencils on the fimo.
2. Lightly grease the rolling-pin. Roll out the fimo in the main colour to make the fin shapes, (A), to 3mm (⅛in) thickness. Lay the acetate stencil over it and cut out the shape twice. Note that the two shapes, when finished, must curve in opposite directions.
3. Make two holes in the top of each piece where shown on the diagram with florist's wire.
4. From the same piece of fimo cut out the shapes (B). Make three holes in them with florist's wire where shown on the diagram.
5. Roll out a contrasting colour of fimo to 3mm (⅛in) thickness to make the shapes (C) and (D). Make one hole through the centre of (C) and of (D), from top to bottom, with florist's wire.
6. From the sheet of acetate film make a straight edge 6mm (¼in) wide and about 7cm (3in) long.
7. Roll out more of the contrasting colour to approximately 1mm thick and lay it along the straight edge you have made.
8. Cut out the small fin bones from this piece, making them 6mm (¼in) long and about 2mm (1/10in) wide. Lift them with the point of your modelling knife and lay them on the fin base, with three rows of four, one row of three, one row of two and one of one, as shown in the diagram.
9. Make three little balls of the contrast colour used for the small fin bones and position them along the top edge of the fin, as shown in the diagram. Make a small hole in each with the point of a pencil, so the main colour shows through.

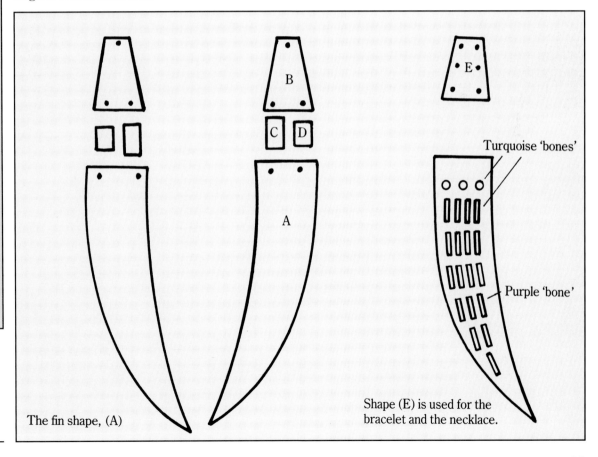

The fin shape, (A)

Shape (E) is used for the bracelet and the necklace.

Turquoise 'bones'

Purple 'bone'

10. Rub over the rolling-pin with an oiled cloth, then roll it over the fin shape, pressing the little 'bones' lightly in place.
11. Turn the fin shape over and repeat steps 8, 9 and 10.
12. If the fin has gone out of shape, replace the stencil over it and trim it round.
13. Bake all the shapes in the oven for twenty to thirty minutes at 130°C (265°F). When baked they will be slightly darker in colour. You should not be able to mark them with your fingernail. If you bake them for too long they will go bubbly and give off fumes. These fumes can be harmful and should not be inhaled. Varnish the shapes when they have cooled.
14. When the varnish is dry, thread a length of embroidery silk through each of the two lower holes in (B) and knot each piece. Thread one piece through (C) and tie another knot. Thread the other piece through (D) and tie another knot.
15. Position the fin piece below the three 'bones' you have just tied together. Tie the thread from the smaller 'bone', (D), to the side towards which the fin curves. Tie the thread from the larger 'bone', (C), to its other side.
16. Finally, slot a large jump ring through the hole in the top part of 'bone' (B). Slot a medium jump ring through the large ring, and finally a piece of kidney wire (the earring clasp) through the medium jump ring.

Bracelet

In addition to the materials listed at the beginning of this section, you will need a metal medium-sized pastry cutter.

1. Make the two fin shapes, (A), as for the earrings.
2. Make one wedge shape, (E), 2mm (¹⁄₁₀in) thick, in the contrast colour.
3. Make three holes through the wide end of each fin with the point of a pencil, and one at the narrow end.
4. Make three holes down each side of the wedge shape.
5. Wrap each fin shape round a metal medium-sized pastry cutter before baking. This will result in a curved shape to fit round the wrist.
6. Bake the shapes as before, and varnish when cool.
7. Lace together the wedge-shaped 'bone' and the wide ends of the two fin pieces as shown in the diagram, and tie the loose threads together at the back.
8. Attach two medium jump rings to each narrow end of the fin shapes. Attach a clasp to one of the jump rings. If you don't want to use a clasp, you can string between twenty and twenty-five small glass beads on the nylon line and attach each end of the line to the jump rings.

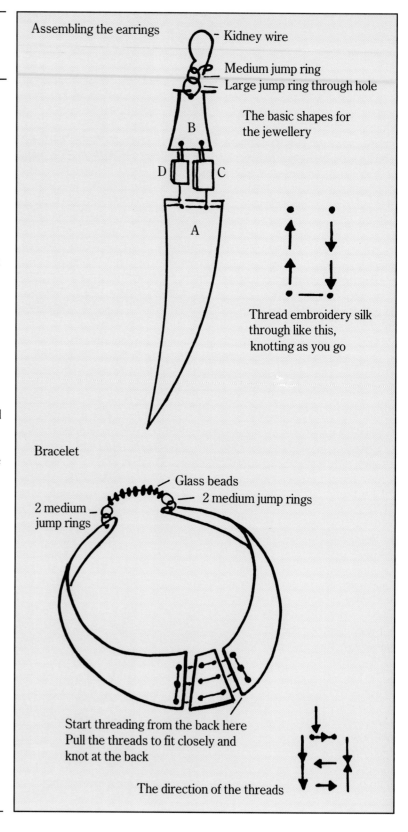

Necklace

1. Make the fin and wedge shapes, (A) and (E), as for the earrings and bracelet, but do not curve the fin shapes when baking. Otherwise bake and varnish as before.
2. Take about a quarter of a block of fimo in the two contrasting colours and roll each into a long sausage shape with your fingers, each about 5mm (¼in) thick and 36cm (14in) long.
3. Twist the two strips together. Fold them in half, laying the two halves side by side and run a lightly greased rolling-pin over them. Then cut them into strips, so that each strip, when rolled into a ball, will be about the size of a pea. Make a few test cuts first until you are sure you have got the size right. You need about fifty pea-sized balls altogether.
4. Push a piece of florist's wire through the centre of each ball to make a hole through it. You can bake the beads on the wire.
5. When the beads are baked and cool, varnish them.
6. Remove the florist's wire from the beads.
7. Attach two medium-sized jump rings to the thin end of each of the fin shapes with about 15-20cm (6-8in) of nylon line. Then thread between ten and twenty small glass beads on each line.
8. Add approximately twenty-five of the fimo beads you have made to each side.
9. Finish the string of beads with a couple more glass beads on each side, then tie the line to a large jump ring each side.
10. Fix the clasp to one of the jump rings.

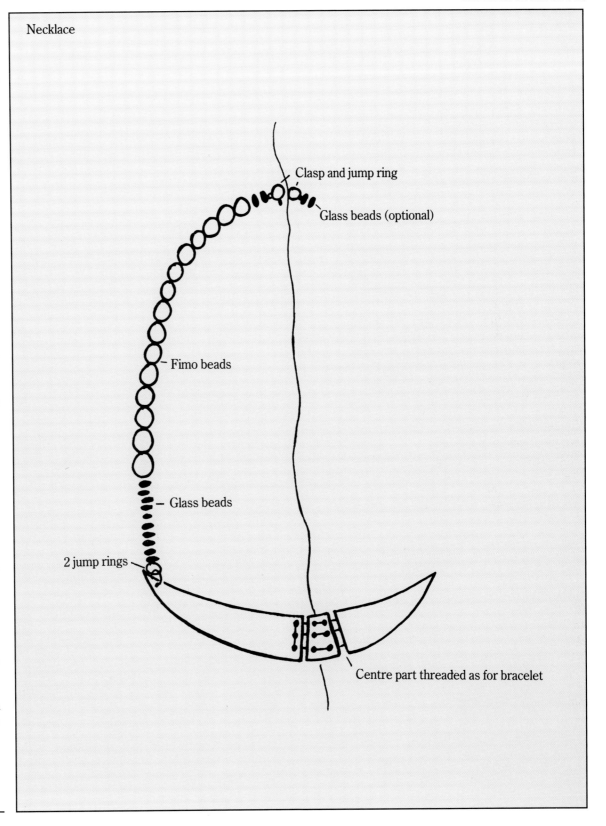

Necklace

On the Button

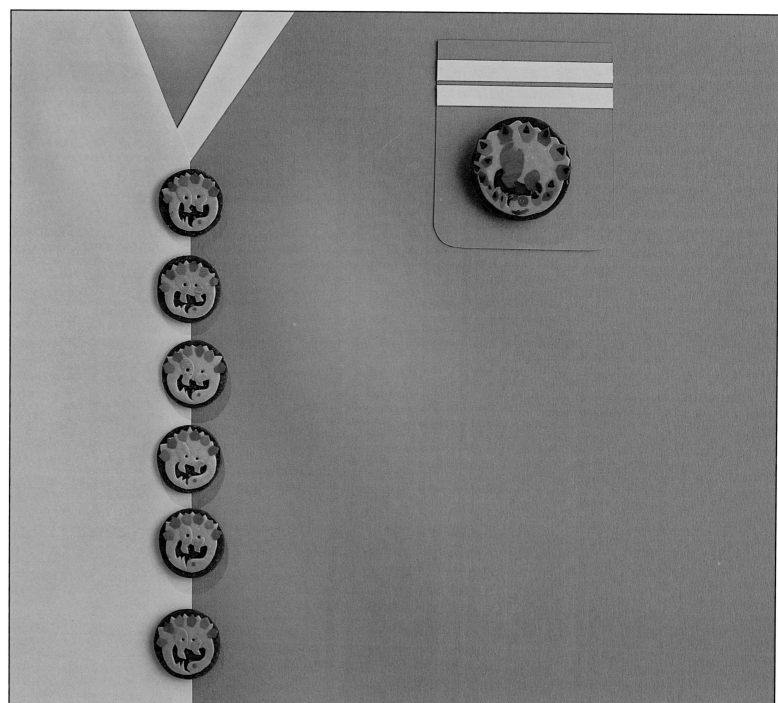

ON THE BUTTON

Fimo is also used to make these jolly stegosaurus buttons. A larger version of the button makes an attractive brooch.

Materials

Fimo in green, yellow, orange, pink and blue
Modelling knife
Sheet of acetate film or thin, smooth card
A pencil
Tracing paper
Fastening and glue for brooch
Rolling-pin
Cooking oil
Darning needle

Method

1. Trace off the designs and make acetate or card stencils of them. Acetate is better because it is easier to peel off the fimo, and also because you can see through it. Wipe it lightly with a cloth greased with cooking oil before putting the stencils on the fimo.
2. Lightly grease the rolling-pin. Roll out the green fimo to make the button base, approximately 4mm (3/16in) thick.
3. Roll out the yellow fimo. Lay the stegosaurus stencil on it and cut it out with a sharp craft knife.
4. Roll out the orange fimo. Lay the leg stencils on it and cut them out with a sharp craft knife.
5. Roll out the pink fimo. Lay the scales stencils on it and cut out 3 of each shape. Also cut out 3 orange scales in the large size.
6. Make tiny balls of blue fimo for the eyes.
7. Lay the yellow stegosaurus body on the green base.
8. Position the legs.
9. Position the pink and orange scales, with the largest scales in the centre of the back and the smaller ones towards the head and tail.
10. Lay the acetate or card over the whole button and smooth the button flat with the thumb and fingers.
11. Position the eyes.
12. Make two holes in the buttons with a darning needle, to make the buttonholes.
13. Bake in the oven for twenty to thirty minutes at 130°C (265°F). When baked they will be slightly darker in colour. You should not be able to mark them with your fingernail. If you bake them for too long they will go bubbly and give off fumes. These fumes can be harmful and should not be inhaled. When the buttons have cooled, varnish them and leave to dry.
14. The brooch is made in the same way, only larger, with more scales and extra specks of colour added to the scales. The fastening is glued on the back with a strong glue.

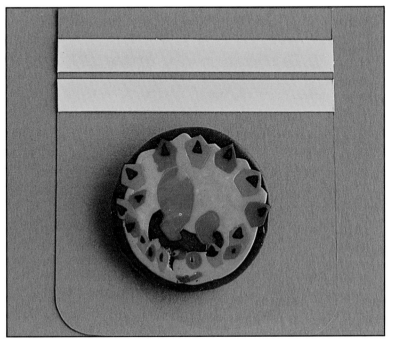

ON THE BUTTON

Primeval Painting: Silk

PRIMEVAL PAINTING : SILK

This scarf and picture are painted on pure silk – which is not as difficult as you might think.

Materials

0.5m of 90cm wide (½yd of 36in) white silk (scarf)
26 x 28cm (10 x 11in) white silk (picture)
Silk paints
Gutta
Adjustable wooden frame or old wooden picture frame
Silk pins or masking tape
A soft painting brush

Before you start

Silk paints can be bought in most handicrafts shops. They may be either water- or alcohol-based: each is used in the same way but the methods of diluting and fixing differ. You should refer to the manufacturer's instructions for how to do this. Silk paints can be mixed in the same way as other paints; they may also be diluted to give less depth of colour.

Gutta is a blocking agent of glue-like consistency, which is used to separate the areas of colour. It is applied by means of a special applicator, which may be used on its own or with a nib. It can be applied clear, or, for a special effect, have silver, black or gold glass paint colouring added to it.

The silk scarf design, which is assembled 1, 2.

1   1cm on the grid equals 2cm

Method

1. Wash any dressing or grease from the silk in hand-warm water using a gentle detergent. Roll the fabric in a clean towel to soak up the excess water, then iron it with a cool iron while still damp.
2. Square up and copy the design shown here, or draw your own on a piece of paper. Use thick black lines so you can see them through the silk.
3. Stretch the silk onto the frame and pin it in place. Special three-pronged flat-sided pins called silk pins are best, as they will not snag the silk. For small areas, such as the square picture, masking tape may be used instead. Work from the centre of one side to the edge, and then from the centre to the other edge.
4. Slip the design under the silk and tape it to the edges. Trace over the design with the gutta, making sure there are no gaps, or the silk paint will bleed through. Make sure the gutta penetrates right through to the back of the silk. Check that the gutta is thick enough by holding the silk up to the light, and if necessary add more.

To avoid air bubbles in the gutta, up-end the applicator vertically at regular intervals. Clean the applicator and nozzle immediately after use with white spirit to prevent them getting blocked. If the gutta is too thick, add a few drops of gutta solvent. Make sure you do not add too much or the gutta will run and not act as a barrier to the paints.
5. Leave the gutta to dry for about an hour. The process can be speeded up with a hair dryer.
6. Apply the silk paints with a good quality brush. This is done not so much by painting as by allowing the paint to diffuse off the brush onto the silk up to the edges of the gutta. Rinse the brush between colours to eliminate any muddiness.
7. When the silk painting is finished and dry, remove it from the frame and fix the colours according to the manufacturer's instructions.

2   1cm on the grid equals 2cm

PRIMEVAL PAINTING : SILK

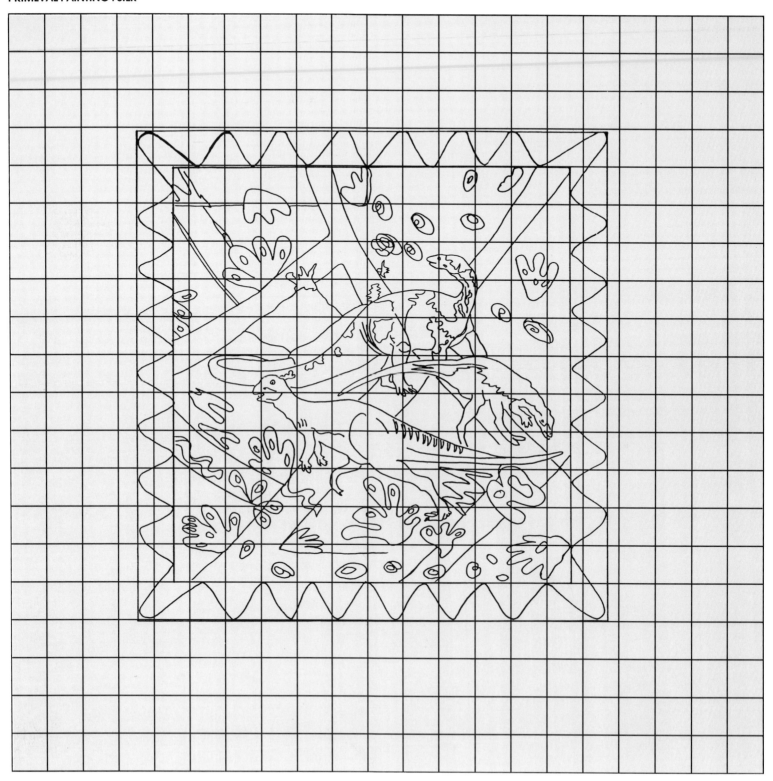

The silk square design   1cm on grid equals 2cm.

DINOSAURS TO PLAY WITH

Tyrannosaurus Tex

This gritty, rough-hewn, cigar-smoking Texan tycoon is a walking puppet.

Materials

21 x 30cm of 10mm (8¼ x 11⅝in of ⅜in) plywood
2 small blocks of timber, 45 x 25 x 19mm (1¾ x 1 x ¾in)
5 small blocks of timber, 50 x 12.5 x 12.5mm (2 x ½ x ½in)
1 strip of wood 275 x 12 x 5mm (10¾ x ½ x ¼in)
1 strip of wood 180 x 12 x 5mm (7⅛ x ½ x ¼in)
1 leather or cord bootlace
1 no. 8 countersunk self-drilling screw, 25mm (1in) long
1 fret-saw or electric skill saw
Enamel paints: blue, orange, buff, black, white, brown, red
Paint brushes
Spool of nylon fishing line
Hand-drill
3mm (⅛in) drill bit
1.5mm (1/16in) drill bit
Medium sandpaper

Method

1. Trace the body and head shapes from the diagram, size up on graph paper, and transfer to the sheet of plywood. Then cut out the shapes with a fret-saw or skill saw.
2. Drill the body where shown for the leg attachments with the 3mm (⅛in) drill bit.
3. Drill the neck and head where shown for the neck cord attachment with the 3mm (⅛in) drill bit.
4. Cut a 50mm (2in) length of bootlace and insert it into the neck and head, leaving 25mm (1in) of neck cord between them.
5. Drill the hat at the point shown with the 1.5mm (1/16in) drill bit.
6. Drill the back at the point shown with the 1.5mm (1/16in) drill bit.
7. Drill the 275mm (10¾in) strip of wood and the 180mm (7⅛in) strip at each of their ends with the 1.5mm (1/16in) drill bit.
8. Place the longer strip on top of the shorter strip to form a cross. It should be placed in the centre of the short strip, and should measure 100mm (4in) from the centre point to its tip.
9. Screw the two strips together from above with the countersunk self-drilling screw.
10. Take the five small blocks of wood and drill them lengthways down the centre with the 3mm (⅛in) drill bit to make five tubes.
11. Cut one of the drilled blocks in half along the 50mm (2in) length and sandpaper them into two bead shapes by rounding off all the edges, to form the lower parts of the legs above the boots.
12. Cut two other drilled blocks across to make one block 31mm (1¼in) long, and the other 19mm (¾in) long. Discard the 19mm (¾in) pieces.
13. Sandpaper the 31mm (1¼in) blocks into bead shapes to make the upper parts of the legs above the boots.
14. Take the remaining two 50mm (2in) long blocks and sandpaper them into tube shapes to make the upper parts of the boots.
15. Take the other two blocks, 45 x 25 x 19mm (1¾ x 1 x ¾in) and using the fret-saw and sandpaper cut and sand them into foot shapes. Drill into the top of each foot through to the instep with the 3mm (⅛in) drill bit.
16. Glue the tops of the boots onto the foot parts.
17. Prime all parts to be painted. Paint the head, body, and short and long leg beads blue. Leave to dry before painting the under body orange. Paint the boots and hat buff, the cigar brown with a red tip. Paint the frame of the sunglasses white, and leave to dry before painting the lenses black. Paint the line of the mouth black, and leave to dry before painting two lines of jagged white teeth. Thread a length of bootlace from the foot of one boot, up through the leg, the short bead and the long bead, then through the under part of the body, and back down the other leg, passing through the long bead, the short bead, the leg of the boot and finally out through the foot part. Tie a knot under each foot, leaving the beads a bit loose.
18. Put a drop of glue into the top of each boot leg with the foot pointing forwards to prevent the foot rotating.
19. Cut four lengths of nylon fishing line, two 50cm (20in) long and two 45cm (18in) long.
20. Tie one piece of the 50cm (20in) line to each leg between the long bead and the short bead, and then to each end of the short strip of the crossbar. The finished length of each piece of line should be approximately 40cm (16in). Don't cut off the end yet.

Tyrannosaurus Tex is shown on the previous page.

21. Tie one piece of the 45cm (18in) line to the hole in the hat, and the other end of it to the front hole on the long piece of the crossbar. The finished length of this line should be approximately 20cm (8in). Don't cut off the end yet.

22. Tie the other piece of the 45cm (18in) line to the hole in the top of the back, and the other end of it to the back hole on the long piece of the crossbar. The finished length of this line should be approximately 33cm (13in) long.

The lengths may need to be adjusted to balance the puppet, hence the adequate line lengths given. You should only cut off the extra lengths when you are sure the puppet is properly balanced, and that it can be made to walk.

23. To make Tex walk, hold the longer crossbar horizontally behind the shorter bar, with the finger and thumb in approximately the centre of the bar. Rock the bar from side to side and move forwards, and Tex will walk forwards, nodding his head.

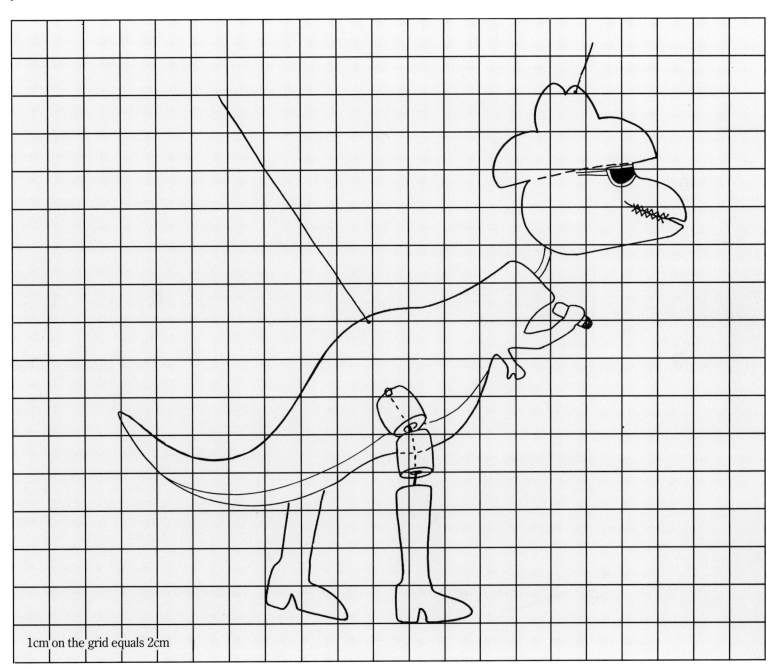

1cm on the grid equals 2cm

TYRANNOSAURUS TEX

Assembly of boot
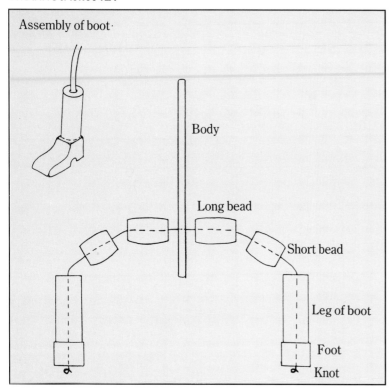

The attachment of the lines
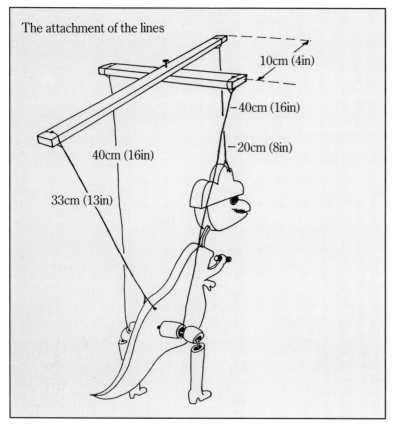

Rocking motion makes Tex walk
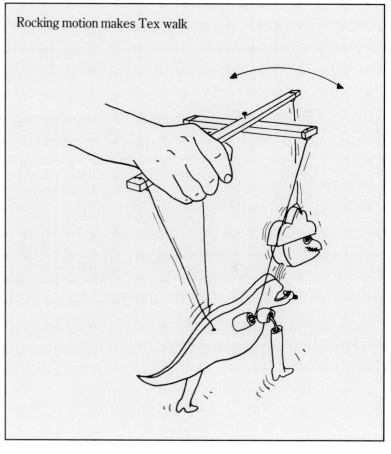

Katie Plateosaurus * Bethia Polacanthus

Katie Plateosaurus

Katie is based on one of the oldest plant-eating dinosaurs. The soles of her feet and the underneath of her chin are white, and she is decorated with small gold beads; white, gold, pink and blue sequins; small clear beads and white bugle beads.

Materials

A 23cm (9in) square of dark blue felt
A scrap of white felt
Dark blue sewing thread
Sequins and small beads
0.5m (½yd) long 2.5cm (1in) wide blue ribbon
225g (8oz) polyester stuffing

Method

1. Scale up the pattern.
2. Cut out two main body pieces, and two underbody pieces from the blue felt.
3. Stitch on the sequins and beads as indicated on the pattern, using a white sequin and a small blue bead to make each eye.
4. Using an over-stitch, sew the underbody to the main body on both sides, (A) to (C) and (D) to (B), on the wrong side.
5. Cut out two sole pieces from the white felt.
6. Sew the sole pieces into the legs, from (C) to (D) and back to (C), matching these points.
7. Cut out the gusset piece from the white felt.
8. Sew the gusset piece under the chin, from (E) to (F) and back to (E), matching these points.
9. Stitch along the whole of the top body seam from (E) to (G), then turn the dinosaur right side out.
10. Cut out the four arm pieces from blue felt.
11. Stitch two arm pieces together, leaving an opening to turn right side out, and repeat with other arm.
12. Stuff the arms and stitch up the openings.
13. Stitch both arms onto the body where indicated on the pattern.
14. Stuff the head, legs and tail firmly.
15. Stitch up the seam on the underbody 3cm (1½in) at a time, stuffing and shaping the body as you go.
16. Finally, tie the ribbon round Katie's neck.

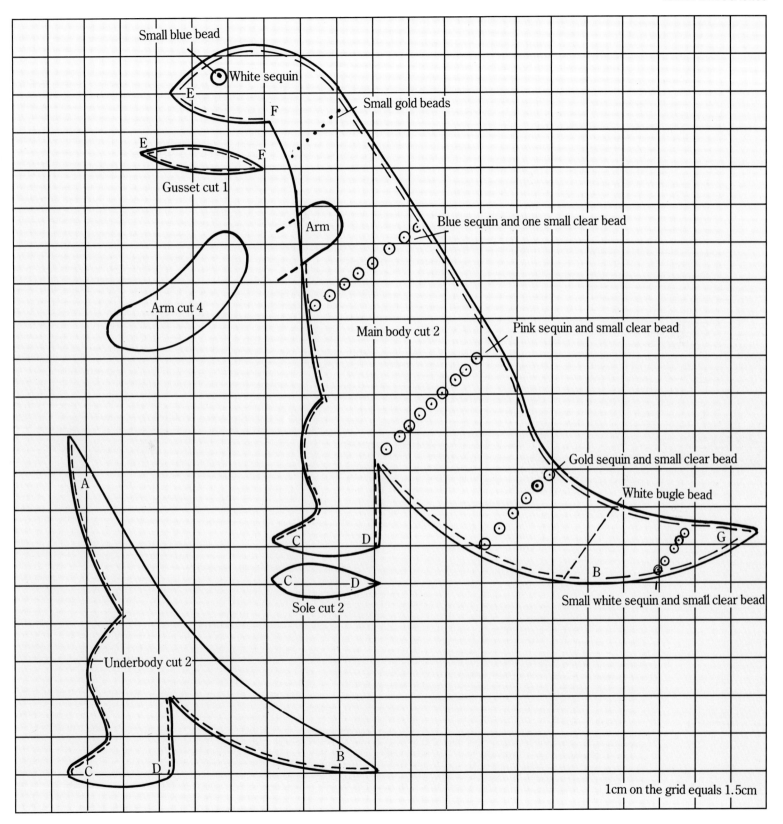

Bethia Polacanthus

Bethia is based on an armoured dinosaur which lived about 120 million years ago. Pink sequins and dark beads are used for the decoration; the scales are made from french knots sewn in matching embroidery cotton with a small dark bead sewn into each knot.

Materials

A 46cm (18in) square of pink felt
A 23cm (9in) square of purple felt
Pink and purple sewing thread
Pink embroidery cotton
Sequins and small beads
0.5m (½yd) 2.5cm (1in) wide pink ribbon
225g (8oz) polyester stuffing

Method

1. Scale up the pattern.
2. Cut out the two body pieces and two gusset pieces in pink felt.
3. Stitch the body pieces together along the top from (C) to (B), using an over-stitch.
4. Stitch the two gusset pieces onto the lower part of the body from (A) to (B) on both sides.
5. Turn the dinosaur right side out.
6. From the purple felt cut one large circle, one medium circle and four small circles. Cut each circle into quarters.
7. These quarter circles make the spikes. You need four large spikes, two medium spikes and fourteen small spikes, and each is made by forming a cone with the quarter circle and stitching the seam from point to base.
8. When you have made the spikes, stuff each one and stitch onto the dinosaur body where indicated on the pattern.
9. Sew a large white sequin and a small dark bead where indicated on the pattern to make each eye.
10. Sew sequins with small dark beads in the centre where indicated on the pattern.
11. Sew the french knots in pink embroidery thread, finishing with a small dark bead where indicated on the pattern, just above the tail.
12. Cut out eight leg pieces from pink felt.
13. Stitch the leg pieces together in pairs, (D) to (E), to make four legs. Turn each leg right side out.
14. Cut out four pieces of back paw and four pieces of front paw in purple felt.
15. Stitch the paw pieces together in pairs to make two back paws and two front paws. Leave the stitching on the right side.
16. Stuff each leg.
17. Stitch a paw onto each leg.
18. Stitch each leg onto the body where indicated on the pattern.
19. Stitch the under seam of the gusset from (D) to (B), leaving the opening for the stuffing.
20. Stuff the head and tail firmly.
21. Stuff the rest of the body, shaping it as you go.
22. Sew up the opening.
23. Finally, tie the ribbon round Bethia's neck.

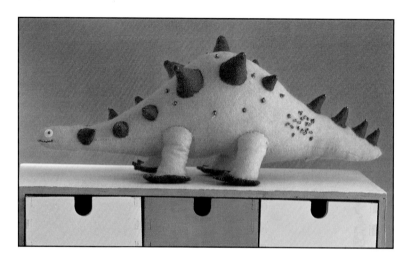

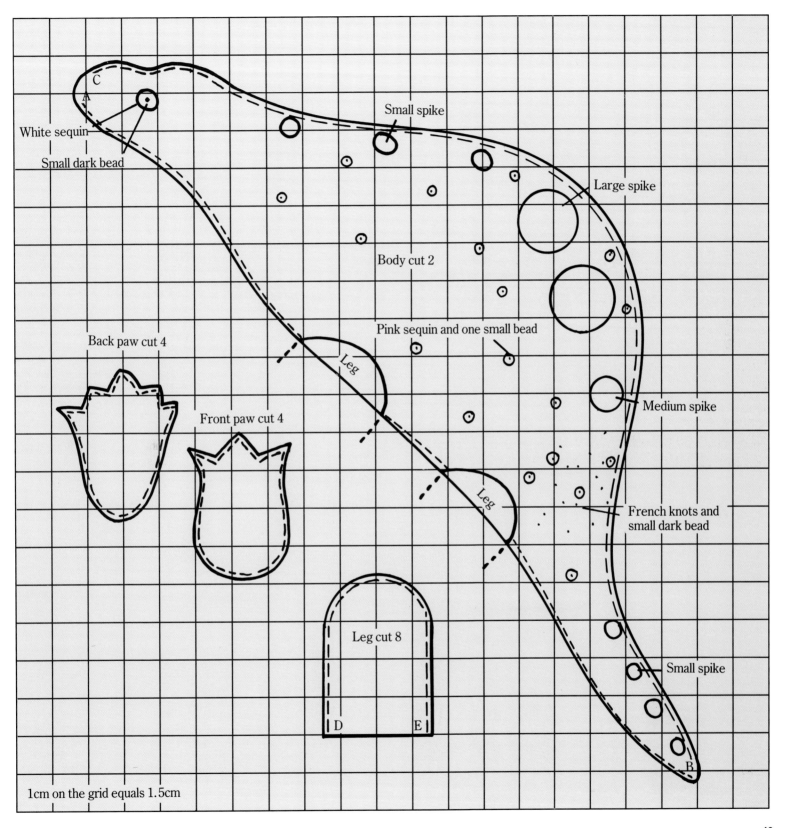

BETHIA POLACANTHUS

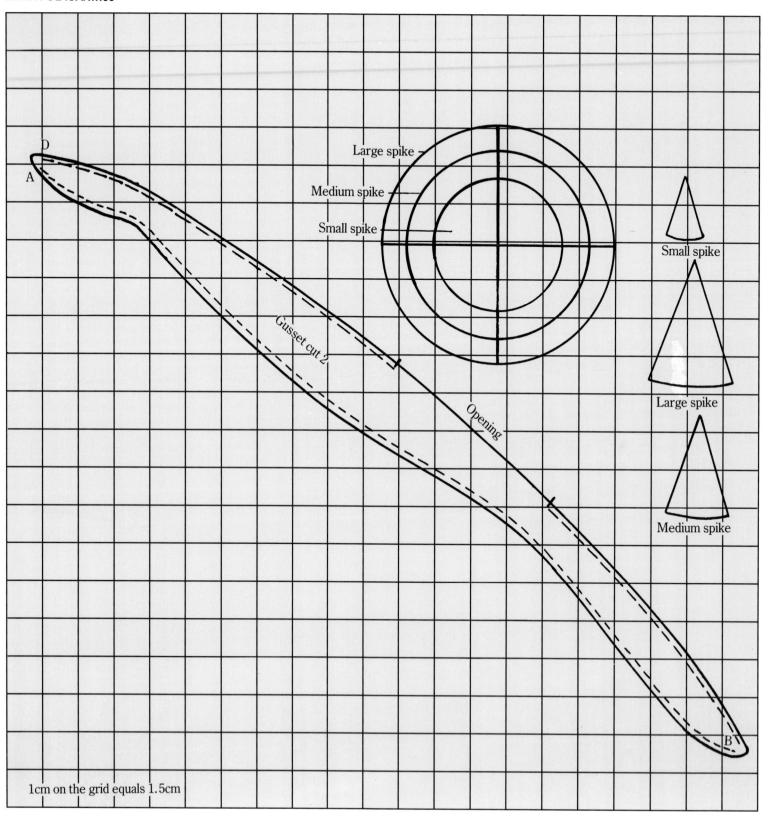

1cm on the grid equals 1.5cm

Monster Mobile

MONSTER MOBILE

Brightly coloured felt shapes make up this monster mobile. It can be decorated with beads or buttons, but if there is any likelihood of the baby being able to grab hold of the mobile, it is better to leave it undecorated. It can be made and hung as described below, or the shapes could be strung from a line of shirring elastic and tied horizontally across a pram or buggy.

Materials

Five 26cm (10in) squares of felt in green, pink, purple, yellow and blue
0.5m (½yd) of 56g (2oz) wadding (optional)
Needle and thread
51cm (20in) fine dowelling
183cm (72in) nylon thread
Glue
Scissors or pinking shears
Paper and pencil
Dressmaker's pins

Method

1. Trace the patterns for the dinosaurs on pages 48 and 49 to make your own paper patterns.
2. Pin the patterns onto the felt and cut out the shapes. Cut two of each to make two sides. If you like, they may be padded out with wadding. If you want to do this, cut one of each shape in the wadding.
3. If you wish to decorate the shapes, do so now.
4. Cut the nylon thread into eight lengths of approximately 23cm (9in). Put a dab of glue on the centre of the back of each dinosaur and sandwich the end of the thread between the two shapes.
5. Sew the two shapes together with a running stitch, sewing as close to the edge as possible, including the wadding if you are using it. Alternatively, the shapes may be glued together.
6. Cut the dowelling into three pieces, one 25cm (10in) long, and two 13cm (5in) long.
7. Tie one dinosaur to the centre of the longest piece of dowelling with the nylon thread, and tie two pieces of thread from the ends of this piece from which to suspend the two shorter pieces of dowelling.
8. Tie the two threads to the centres of the shorter pieces of dowelling.
9. Tie two dinosaurs to the ends of each of the shorter pieces of dowelling.
10. Lastly tie one thread from the centre of the longer piece of dowelling, from which the whole mobile will be suspended.
11. You will need to adjust the lengths of the threads to achieve the effect you desire. The simplest way to do this is to wind the thread round the dowelling.

MONSTER MOBILE

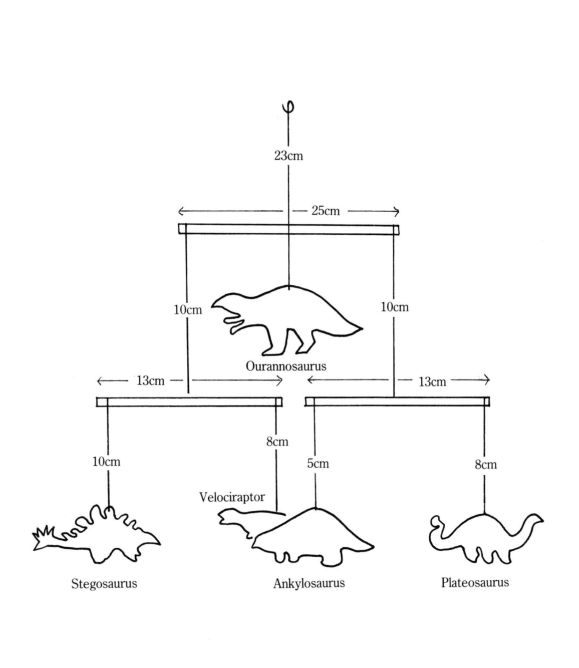

MONSTER MOBILE

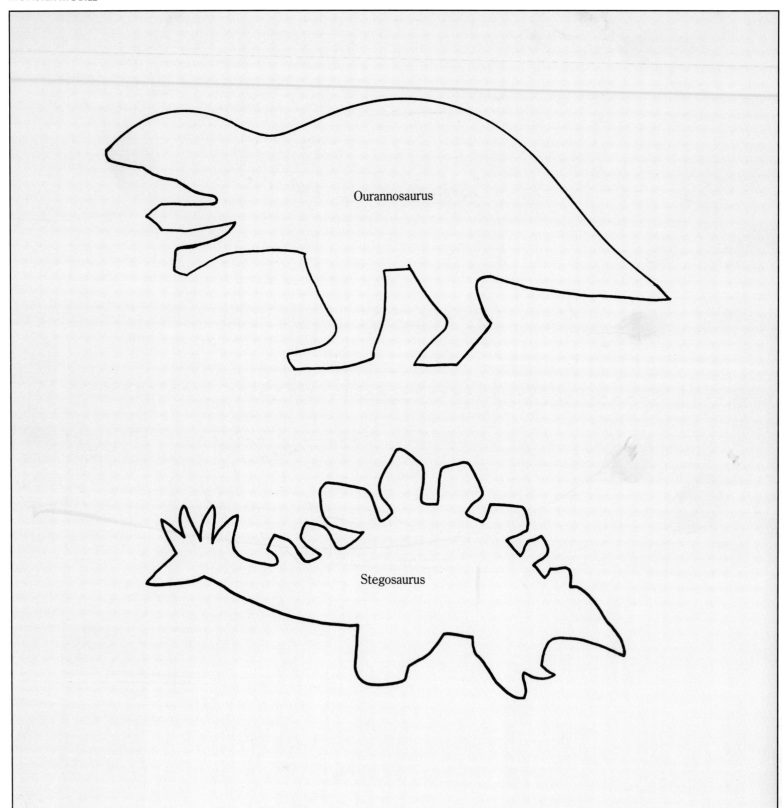

MONSTER MOBILE

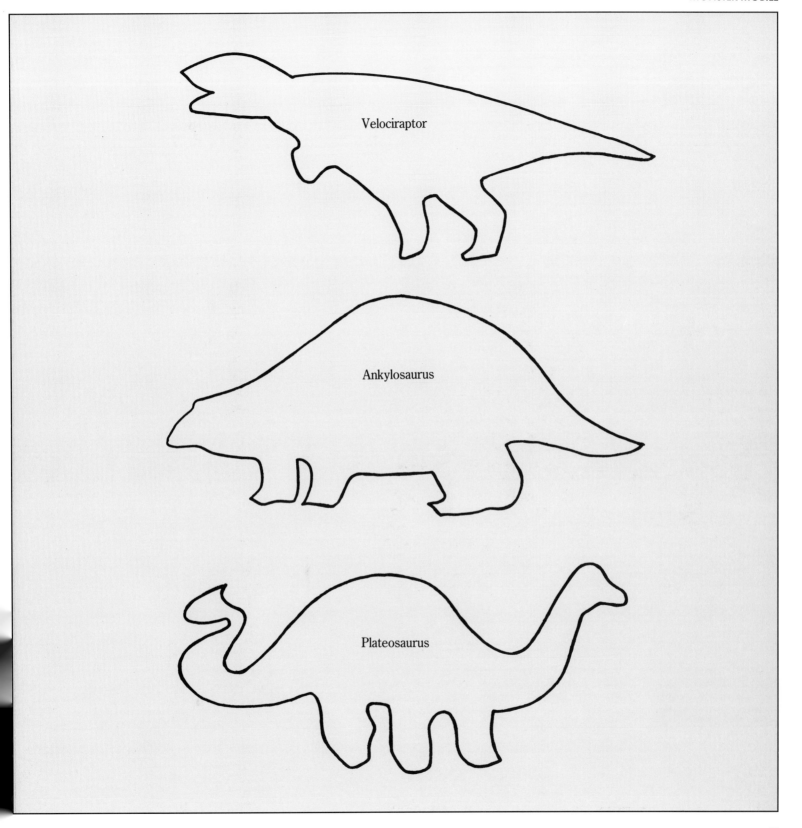

Pterodactyl Gliders

PTERODACTYL GLIDERS

Children will love playing with these paper gliders, which swoop and fly around the room. Diagrams are given for making one example, but it is fun to experiment and create unique models of your own, once you get the hang of the aero-dynamics.

Materials

Sheet of stiff paper in main colour measuring approximately 36 x 20cm (14 x 8in)
Sheet of stiff paper in a contrast colour measuring approximately 30 x 8cm (11 x 3in)
Pencil
Scissors
Paper glue
Black felt pen

Method

1. Scale up the pattern on page 52, and trace it off onto the paper.
2. Cut one shape (A) from the main colour.
3. Cut two shapes of (B), (C) and (D) from the main colour.
4. Draw the eye shapes on the two (D) pieces with the black felt pen.
5. Place the sheet of contrast colour on top of the sheet of main colour. Cut out two shapes of (E) and (F) by laying them on the paper, tracing round them, then flipping them over from the centre line and drawing a second half, or mirror image.
6. Take shape (A), with the tail fin uppermost, and glue a shape (B) on either side of it.
7. Then glue a shape (C) on either side of that, and finally a shape (D) on the outside of each, as shown in the diagram.
8. Glue the wings into position (i) to 1 with the contrast colour facing downwards.
9. Glue the tail wings into position (ii) to 2 with the contrast colour facing downwards.
10. Gently fold the wings into shape, folding the front edge down. The wings should have a slight V shape when viewed from the front. Your pterodactyl should now be ready to fly.

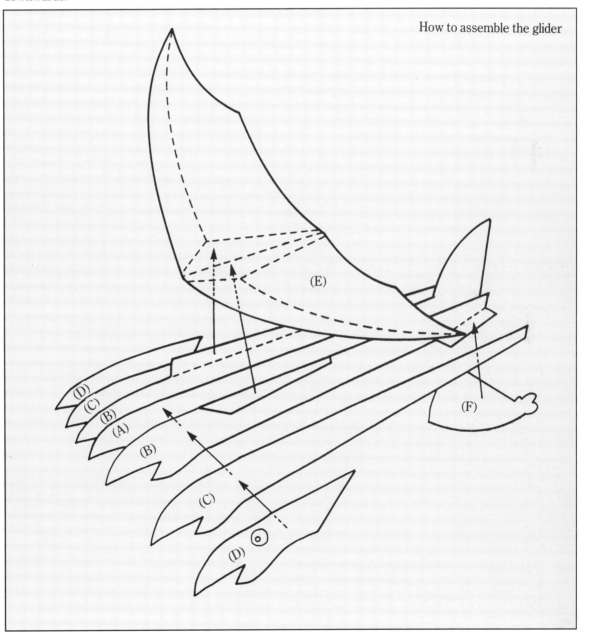

How to assemble the glider

PTERODACTYL GLIDERS

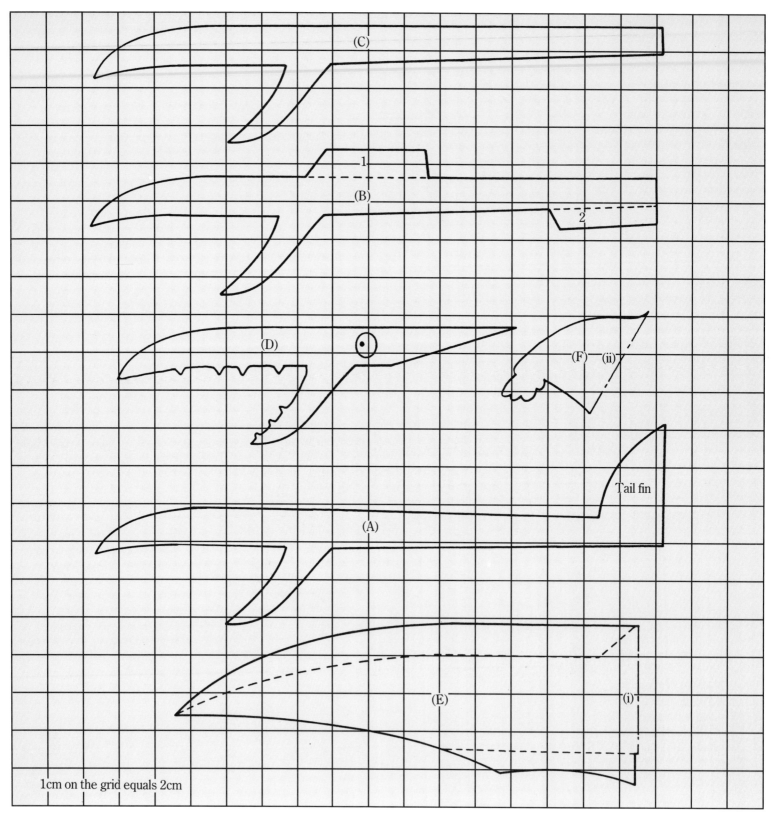

1cm on the grid equals 2cm

DINOSAURS IN THE HOME

Bathtime Behemoth

A friendly protoceratops makes an ideal bath mat.

Materials

1. 30m of 100cm wide (1½yd of 44in) lemon towelling
30cm of 100cm wide (⅓yd of 44in) green towelling
Matching thread
Dressmakers' graph paper
Tracing paper and pencil
Dressmakers' pins

Method

1. Size up the pattern given and add 1cm (½in) seam allowance around the main shape.
2. Cut out the main protoceratops shape twice in lemon towelling.
3. Cut out the underbody, spots, eye and other areas marked green from the green towelling.
4. Using the pattern as a reference, pin all the green shapes in position onto one of the protoceratops-shaped pieces of towelling.
5. Sew all these pieces in place with a running stitch.
6. Using a close zig-zag stitch, go over all the rough edges and the running stitch round each piece being appliquéd in place. If you want a stronger outline, go over the edges a second time.
7. To indicate the curve of the legs, the mouth and the markings round the head, sew a line of satin stitch as shown in the diagram.
8. Pin the top side of the mat, with the appliquéd pieces, to the under side, with right sides facing. Stitch the two pieces together, leaving a gap for turning.
9. Clip any awkward bends and corners, and turn the mat right side out. Slip-stitch the opening closed.
10. To finish, iron flat, then sew a line of top-stitching 0.5cm (½in) from the edge of the mat.

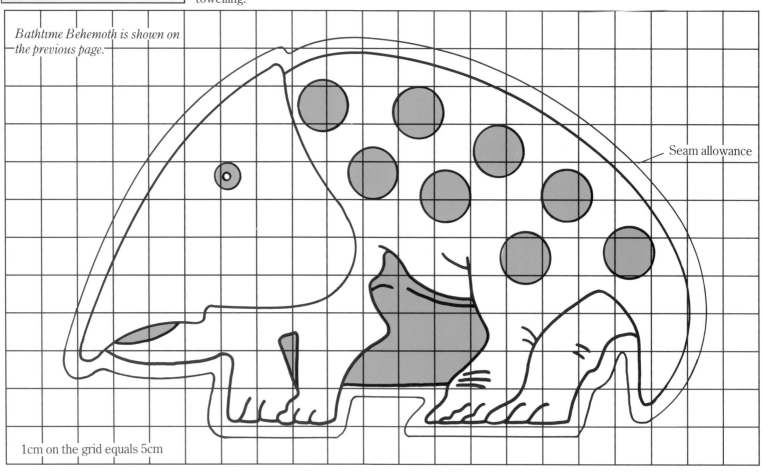

Bathtime Behemoth is shown on the previous page.

1cm on the grid equals 5cm

Bookworms

BOOKWORMS

Dinosaurs make excellent book ends. Two designs are given here, so you can make two the same or one of each.

Materials

30 x 21cm of 10mm (12 x 8in of ⅜in) plywood
15 x 5 x 5cm (6 x 2 x 2in) softwood block
18 x 8.5cm (7⅛ x ¾in) flat plastic sheet, e.g. from ice-cream tub
2.5cm (1in) 2 x 2cm (¾ x ¾in) timber angle fillet
5 no. 8 countersunk self-drilling screws, 25mm (1in) long
General purpose glue
Fret-saw or electric skill saw
Enamel paints: dark blue, bright pink, black, white, green, brown
Paint brushes
Medium sandpaper

Edmontosaurus

Method

1. Trace the dinosaur shape from the diagram and transfer onto the plywood sheet. Cut out the shape with a fret-saw or electric skill saw.
2. Take the softwood block and cut a groove 10mm (⅜in) wide and 15mm (⅝in) deep along its top.
3. Using a rectangular polythene sheet, e.g. the top of a large ice-cream tub or plastic food container, cut a strip 18cm (7⅛in) long and 8.5cm (3¼in) wide. Glue the strip to the underside of the softwood block, positioning it centrally.
4. Using the five countersunk self-drilling screws, screw the plastic to the block.
5. Sandpaper the top edges of the block to a rounded profile. Sandpaper the rest of the block, and the cut-out dinosaur, smooth.
6. Cut a 2.5cm (1in) length from a piece of 2 x 2cm (¾ x ¾in) timber angle fillet to make a 'book'. Sandpaper the book smooth.
7. Prime, and then when dry, paint the book, the outside green, the inside white. When the inside is dry you can add text in black, if you like, such as, 'In the beginning…'.
8. Prime, and then paint, the dinosaur. Do the dark blue first, leaving a little space round the edges for the pink paint to show. When quite dry, paint the pink. Finally add the nose in black, the spectacles and outside of the eye in white, and lastly, after the white paint has dried, the centre of the eye in black.
9. Prime, and then paint the softwood block brown, to represent a sofa.
10. When all the paint is dry, slide the book into the V shape of the 'hands'. Glue it in place with general purpose glue.
11. Finally, slide the dinosaur into place in the groove in the block, with the plastic sheet facing away from the side you want to see.

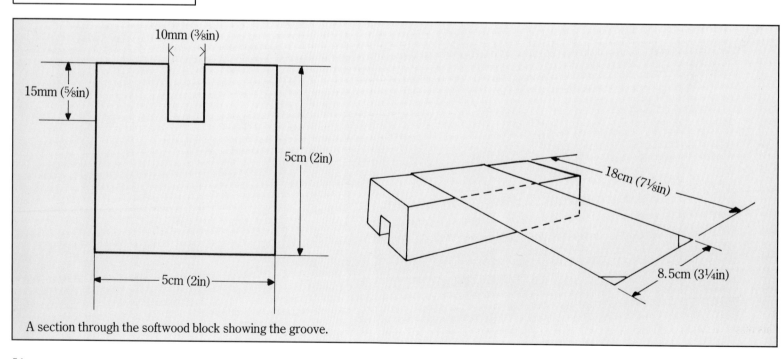

A section through the softwood block showing the groove.

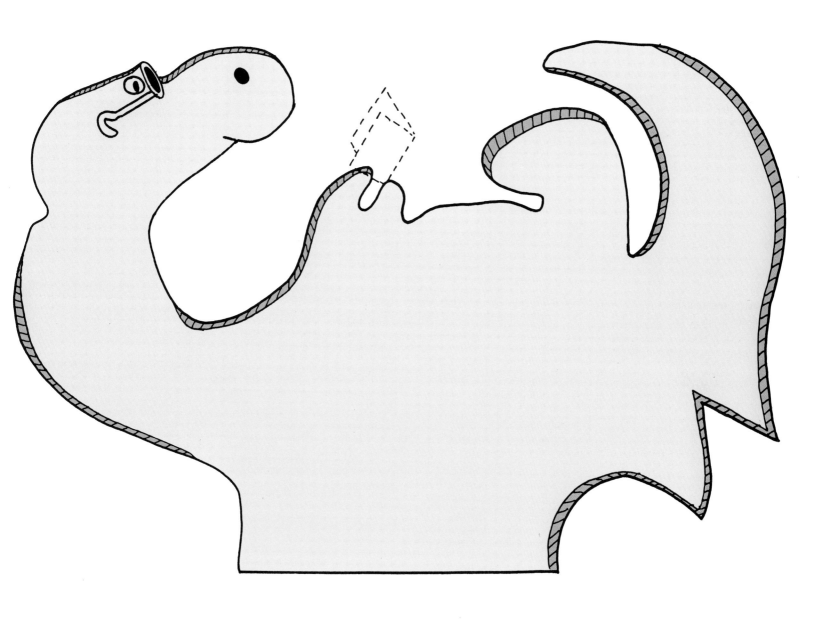

This diagram is actual size.

Apatosaurus

Materials

- 30 x 21cm of 10mm (12 x 8in of ⅜in) plywood
- 15 x 5 x 5cm (6 x 2 x 2in) softwood block
- 18 x 8.5cm (7⅛ x ¾in) flat plastic sheet, e.g. from ice-cream tub
- 2.5cm (1in) 2 x 2cm (¾ x ¾in) timber angle fillet
- 5 no. 8 countersunk self-drilling screws, 25mm (1in) long
- General purpose glue
- Fret-saw or electric skill saw
- Enamel paints: purple, yellow, brown, green, white, black
- Paint brushes
- Medium sandpaper

Method

1. Trace the dinosaur shape from the diagram, size up on graph paper, and transfer to the plywood sheet. Cut out the shape with a fret-saw or electric skill saw.
2. Take the softwood block and cut a groove 10mm (⅜in) wide and 15mm (⅝in) deep along its top.
3. Using a rectangular polythene sheet, e.g. the top of a large ice-cream tub or plastic food container, cut a strip 18cm (7⅛in) long and 8.5cm (3¼in) wide. Glue the strip to the underside of the softwood block, positioning it centrally.
4. Using the five countersunk self-drilling screws, screw the plastic to the block.
5. Sandpaper the top edges of the block to a rounded profile. Sandpaper the rest of the block, and the cut-out dinosaur, smooth.
6. Cut a 2.5cm (1in) length from a piece of 2 x 2cm (¾ x ¾in) timber angle fillet to make a 'book'. Sandpaper the book smooth.
7. Prime, and then when dry, paint the book, the outside green, the inside white. When the inside is dry you can add text in black, if you like, such as, 'The history of the world...'.
8. Prime, and then paint, the dinosaur. Do the purple first, then, when it is quite dry, the yellow. When the yellow is dry, the purple lines can be added. Finally add the white for the spectacles and white of eye, and when that is dry, the black for the centre of the eye and the nose.
9. Prime, and then paint the softwood block brown, to represent a sofa.
10. When all the paint is dry, slide the book into the V shape of the 'hands'. Glue it in place with general purpose glue.
11. Finally, slide the dinosaur into place in the groove in the block, with the plastic sheet facing away from the side you want to see.

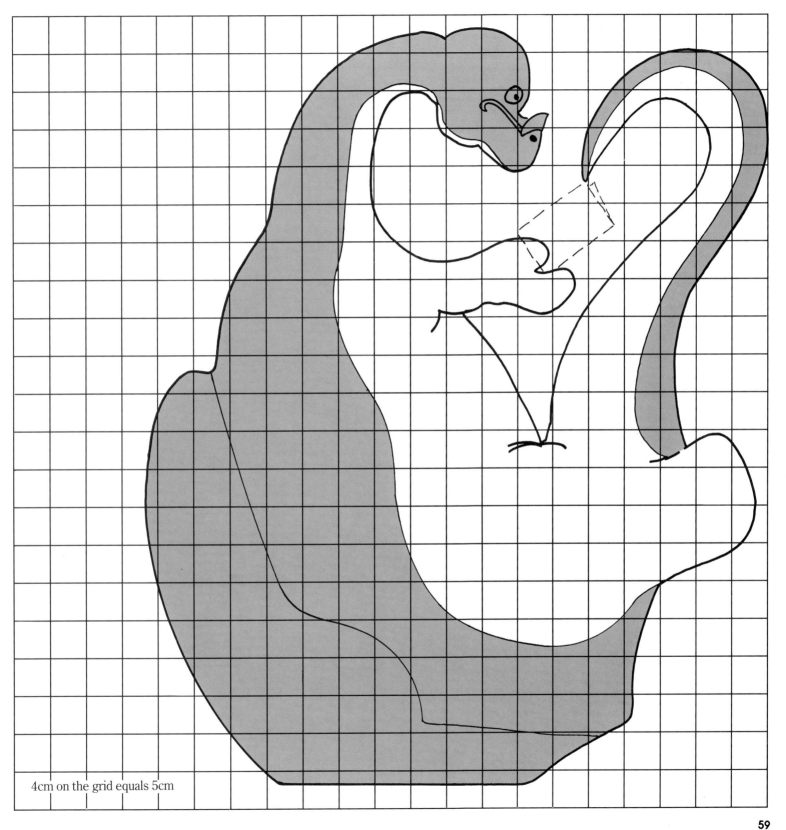

4cm on the grid equals 5cm

Easy Option

This smiling stegosaurus makes a comfortable cushion on which to relax.

Materials

90cm of 120cm wide (1yd of 48in) turquoise cotton
70cm of 120cm wide (¾yd of 48in) orange cotton
50cm of 90cm wide (20in of 36in) wadding
1 38cm (15in) diameter cushion
1 250g (9oz) bag polyester fibre filling
1 25cm (10in) turquoise zip
2 large 'googly' eyes
Black embroidery thread

Method

1. Scale up the pattern.
2. Cut out of the turquoise cotton two body pieces (A); four pairs of legs (B); one top gusset (C); two under-gussets (D).
3. Cut out of the orange cotton four back spikes (E); four tail spikes (F); sixteen toes (G); twelve spots (H) and four feet (I).
4. Cut out of the wadding two back spikes (E) and six spots (H).
5. To make the back spikes, lay two pieces of fabric (E) right sides together, then place one piece of wadding on the wrong side. Pin and then stitch round the shaped edge. Trim the seam and snip into the corners. Turn right side out. Keep the wadding and fabric together at the open edge by pinning and stitching. Repeat to make the other back spike.
6. To make the tail spikes, lay one piece of fabric (F) right sides together, pin and stitch. Trim seam. Turn right side out and stuff. Repeat to make the other tail spikes.
7. To make the spots, lay two pieces of fabric (H) right sides together, then place one piece of wadding on the wrong side. Pin and then stitch all round the edge. Trim the seam and make a small slit in the centre of the circle and turn inside out. Press. Repeat until you have made six spots. Stitch three of each onto the outside of each body piece (A) as shown on the pattern.
8. To make the toes, lay two pieces of fabric (G) right sides together, pin and stitch along the shaped edge. Trim the seam, turn right side out, and press. Repeat to make four pairs of toes.
9. Top-stitch each made toe into position on the right side of each leg.
10. To make the legs, lay the right sides of the fabric together, then pin and stitch round the top part of each leg. Trim the seams, turn right sides out, and press. Make four legs.
11. With right sides together, pin a foot to each leg and stitch, leaving a small gap on the inside of each leg to turn. Trim the seam and turn right sides out. Make four feet.
12. Pin and stitch each leg into position on the body, remembering to put the gaps in the legs on the insides.
13. Pin and stitch the back and tail spikes into position on the body and tail.
14. Position the eyes as shown on the pattern.
15. With a 2cm (¾in) seam, sew the two pieces of under-gusset (D) together, leaving a gap for the zip. Then sew in the zip and trim the seam.
16. Sew together the top and under-gussets at the straight edge. Matching the notches, seam one side of the gusset to the body. Repeat with the other side. Turn inside out through the zip.
17. Stuff the legs, then sew up the openings neatly.
18. Stuff the head and tail.
19. Embroider the mouth as shown on the pattern, with black embroidery thread.
20. Insert the cushion.

EASY OPTION

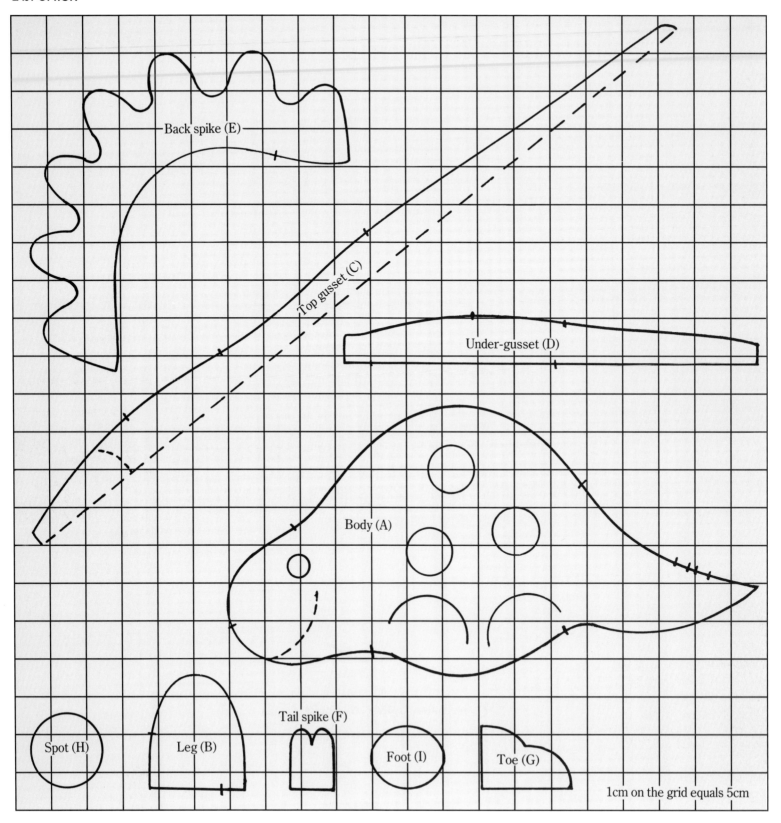

Dimetrodon Draught-stop

DIMETRODON DRAUGHT-STOP

This dimetrodon will be an attractive and useful addition to any chilly house. If he is to be used solely as a draught excluder, then you only need make and sew on two legs, though instructions are given for making four. If only using two legs, don't forget to decide which way you want him to face.

Materials

50cm of 120cm wide (20in of 48in) green cotton
50cm of 120cm wide (20in of 48in) pink cotton
60cm of 120cm wide (24in of 48in) purple cotton
1m of 90cm wide (1yd of 36in) wadding
2 250g (9oz) bags polyester fibre filling
2 large 'googly' eyes
Black embroidery thread

Method

1. Scale up the pattern.
2. Cut out of the purple cotton two top body pieces (A) and four pairs of legs (B).
3. Cut out of the green cotton two back frill pieces (C) and two pairs of feet (D).
4. Cut out of the pink cotton two side body pieces (E) and one underbody (F).
5. Cut out of the wadding four pieces of back frill (C).
6. Place two leg pieces of fabric (B) right sides together and seam around the top part of the leg. Trim the seam, turn right sides out, and press. Repeat to make two or four legs.
7. With right sides together, pin a foot to each leg and stitch, leaving a small gap on the inside of each leg to turn right sides out. Trim the seam and turn.
8. Seam the top body (A) and side body (E) pieces together.
9. Pin and top-stitch the legs into position, remembering to put the gap in the legs on the inside.
10. Position the eyes as shown on the pattern.
11. Place the back frill pieces (C) right sides together, then place two pieces of wadding (C) on each wrong side. Pin and stitch round the shaped edge. Trim the seam, snipping the corners and turn right side out. Keep the wadding and fabric together at the open edge by pinning and stitching before top-stitching down each spine as shown on the diagram.
12. With right sides together, sew the back frill to one side of the top body, matching the notches. Sew the other top body piece to it, right sides together, with the frill sandwiched between and pointing downwards. Seam right along the top body.
13. Seam the side body pieces (E) to the underbody (F) on both sides, leaving a 10cm (4in) gap to turn the dinosaur through.
14. Turn right sides out and stuff the body and legs.
15. Stitch up the gaps neatly.
16. Embroider the mouth as shown on the pattern with black embroidery thread.

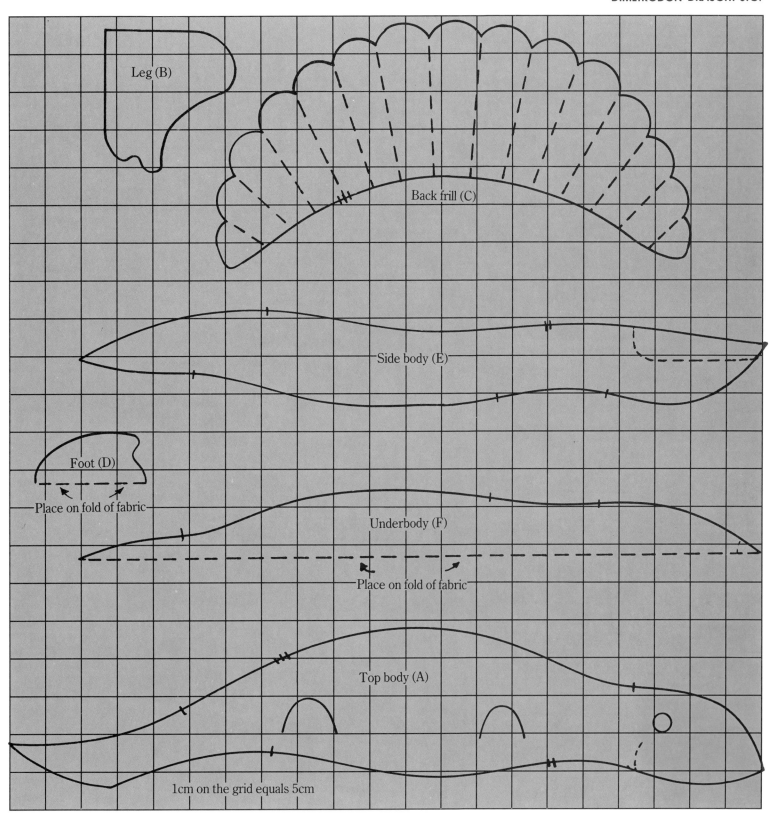

Eggheads

These three cheerful egg-cosies may be hand- or machine-sewn, and will brighten up any breakfast table.

Stegosaurus

Materials

Four pieces of felt in main colour to make body, head, tail and lining
Two pieces of felt in contrast colour for toenails
One piece of felt in contrast colour for spikes
Small amount of stuffing
Black embroidery thread

Method

1. Cut out the two main body shapes and the two lining shapes in the main colour.
2. Cut out the two toenail pieces in a contrasting colour.
3. Cut out the spikes in the toenail colour.
4. Using two strands of black embroidery silk stitch the eyes, mouth and leg line in back-stitch on one main body piece of felt.
5. Sew on the toenails.
6. Sew the two main body pieces together, slipping the spikes in between the top part and stitching through all the layers, 0.5cm (¼in) from the edge in back-stitch, leaving the bottom edge open.
7. Stitch the lining pieces together 0.5cm (¼in) from the edge, leaving the bottom edge open.
8. Push a *little* stuffing into the head and tail of the main body.
9. Slip-stitch the lining into the dinosaur, with wrong sides together.
10. Back-stitch all round the bottom edge 0.5cm (¼in) from the edge.

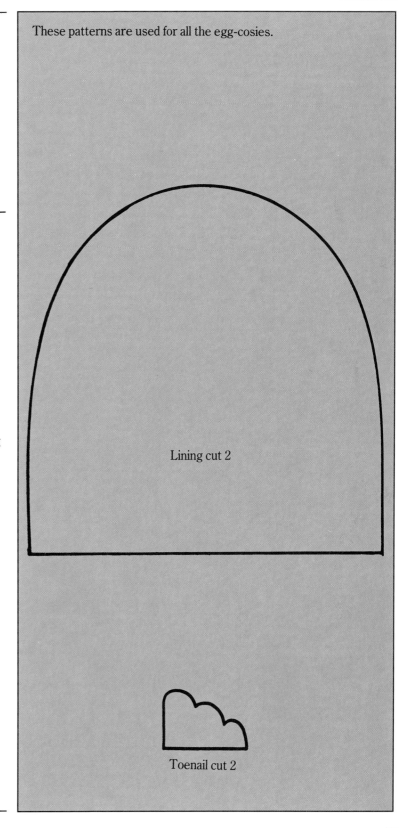

These patterns are used for all the egg-cosies.

Lining cut 2

Toenail cut 2

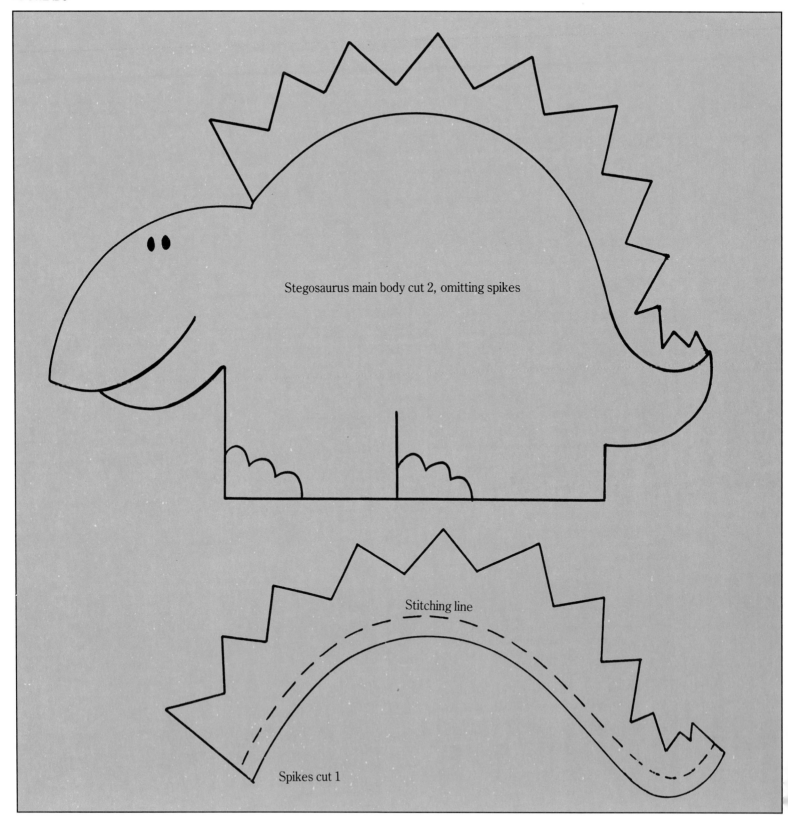

Diplodocus

Materials

Four pieces of felt in main colour to make body, head, tail and lining
Two pieces of felt in contrast colour for toenails
Small amount of stuffing
Black embroidery silk

Method

1. Cut out the two main body shapes and the two lining shapes in the main colour.
2. Cut out the two toenail pieces in a contrasting colour.
3. Using two strands of black embroidery silk stitch the eyes, mouth and leg line in back-stitch on one main body piece of fabric.
4. Sew on the toenails.
5. Sew the two main body pieces together 0.5cm (¼in) from the edge in back-stitch, leaving the bottom edge open.
6. Stitch the lining pieces together 0.5cm (¼in) from the edge leaving the bottom edge open.
7. Push a *little* stuffing into the head and tail of the main body.
8. Slip-stitch the lining into the dinosaur, with wrong sides together.
9. Back-stitch all round the bottom edge 0.5cm (¼in) from the edge.

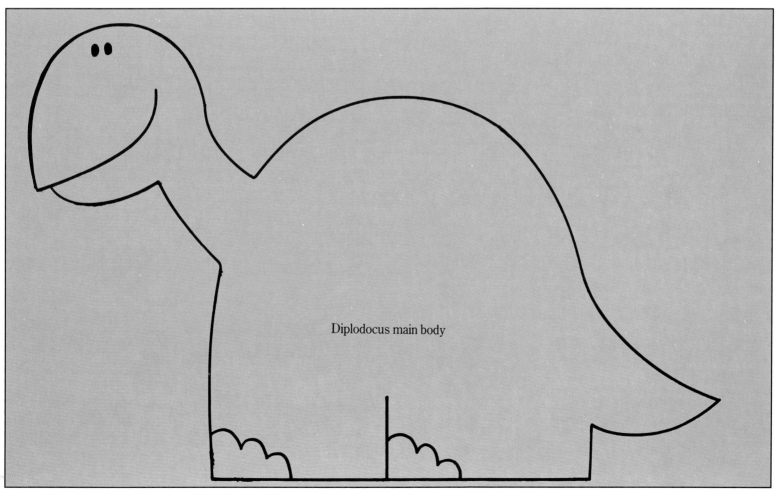

Diplodocus main body

Triceratops

Materials

Four pieces of felt in main colour to make body, head, tail and lining
One extra piece of felt in main colour to make head/collar
Two pieces of felt in contrast colour for toenails
One piece of felt in contrast colour to make three horns
Small amount of stuffing
Black embroidery thread

Method

1. Cut out the two main body shapes and the two lining shapes in the main colour.
2. Cut out the extra head/collar piece in the main colour.
3. Cut out the two toenail pieces in a contrasting colour.
4. Cut out the three horn pieces in the toenail colour.
5. Sew on the toenails.
6. Pin the extra head/collar piece in place over the head on the main body piece. Stitching through both pieces, sew the eyes, nose and mouth lines in back-stitch, using two strands of black embroidery silk, and sew the horn pieces in place. This will attach the extra piece to the main shape.
7. Sew the leg line on the main body piece, in back-stitch.
8. Sew the two main body pieces together 0.5cm (¼in) from the edge in back-stitch, leaving the bottom edges open.
9. Stitch the lining pieces together 0.5cm (¼in) from the edge, leaving the bottom edge open.
10. Push a *little* stuffing into the head and tail of the main body.
11. Slip-stitch the lining into the dinosaur, with wrong sides together.
12. Back-stitch all round the bottom edge 0.5cm (¼in) from the edge.

Cosy Dimetrodon

COSY DIMETRODON

This cheerful tea-cosy would look well on the breakfast table with the egg-cosies on page 66.

Materials

35cm of 90cm wide (14in of 36in) polycotton in main colour

35cm of 90cm wide (14in of 36in) polycotton in a contrast colour

35cm of 90cm wide (14in of 36in) wadding

Sewing thread

Method

1. Size up the pattern and add a 1cm (½in) seam allowance.
2. Cut out two of the main shape, (A), in the main colour, and two in the contrast colour, which will be the lining.
3. Cut out two of the main shape, (A), in the wadding, trimming it to 0.5cm (¼in) smaller than the fabric.
4. From the contrast colour cut out two of each of the fin shape, (B), the tail shape, (C), and the head shape, (D). Cut out one of each of these shapes from the wadding.
5. Pin and tack the head wadding to the wrong side of one of the head shapes, (D). Remove the pins. Then, with right sides together, sew the two pieces of head fabric together, using a small running stitch. All three pieces should then be joined. Turn to the right side.
6. Repeat step 5 with the fin piece and the tail piece.
7. Take the fin piece, and sew lines of straight stitch to divide it up into separate fins, as shown on the pattern. Then go over these lines with a close-set zig-zag stitch. Top-stitch round the top edge of the fins to give greater definition.
8. Embroider the eyes on the head piece using satin stitch, or, alternatively, sew on buttons for the eyes. Embroider the mouth line using satin stitch.
9. Sandwich the fin, head and tail between the right sides of the outer covering, pointing inwards and downwards, and pin in place. Sew round the curved edge, thus joining the two sides of the cosy and securing the fin, head and tail as you go.
10. With right sides together, sew the two pieces of lining together round the curved edge.
11. Sew the two main pieces of wadding together round the curved edge.
12. Place the wadding over the wrong sides of the outer covering, then add the lining with the right side uppermost. Turn under all raw edges and hand-sew the lining to the outside.
13. To prevent the lining and wadding dropping down when the cosy is lifted off the teapot, sew a few stitches along the curved edge catching all three layers together.
14. Finally, turn the tea-cosy to the right side.

COSY DIMETRODON

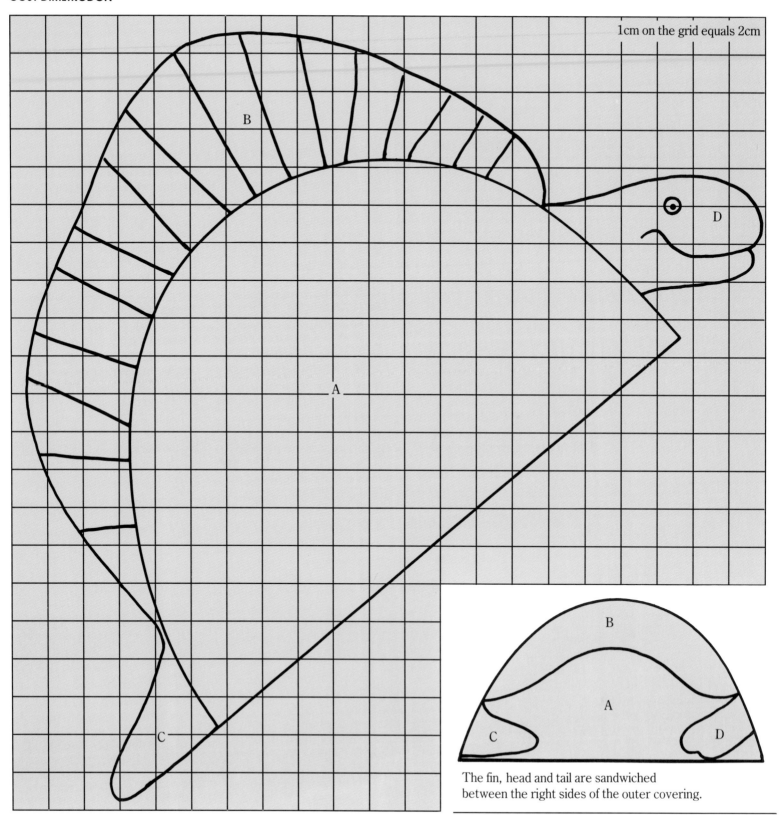

1cm on the grid equals 2cm

The fin, head and tail are sandwiched between the right sides of the outer covering.

Primeval Painting: China

PRIMEVAL PAINTING : CHINA

Painting your own design on china is great fun, and gives lots of scope for creating commemorative plates and presents.

Materials

Plain plates or mugs
Ceramic paints
Chinagraph pencil
White spirit
Paintbrushes
Saucers in which to mix colours

Before you start

The paints used here are for glazed china, which means you can rub off your mistakes, as they don't sink into a non-porous surface. But it also means that although the painted objects make wonderful presents or commemorative plates, they do not stand up to heavy use. It is not advisable, for example, to use a knife and fork on a painted plate, and it should never be put in a dishwasher.

Ceramic paints can be found in most good craft shops. We have used solvent-based air-drying ceramic paints called Ceramic à Froid, which are fixed by leaving to dry in the air for about 24 hours. Other manufacturers use different methods of fixing, so read the instructions carefully before you start.

A semi-opaque finish can be achieved by mixing ceramic paints with stained glass paints.

Before you start painting, make sure you have some old, clean newspapers around on which to stand objects while they are drying. You will also need plenty of white spirit for washing brushes and diluting the paints, if they are solvent-based, and soft cloths – one dipped in white spirit to wipe off mistakes, and one dry one to remove any smears left by the wet one.

After painting, you can protect the painted objects with a coat of clear varnish. Painted objects should be handled with care, and hand-washed in a mild detergent and warm water.

If possible, try to work somewhere free from static, as ceramic paints attract any small particles of fluff which happen to be around.

Method

1. Using a chinagraph pencil, copy the design onto your plate.
2. Paint in the large areas of colour first. If you need to mix a colour, make sure you have mixed enough before you start. Ceramic paints look better slightly dappled, as it is very hard to achieve a flat, even surface over a large surface area. This makes them ideal for painting dinosaurs with their leathery-looking skin.
3. Once you have painted the large areas and they have dried, add the details. Do not overload the brush with paint, but gradually build up the colour.
4. Add a fine black outline if this is needed for fur or fine markings.
5. Leave until all the paint is quite dry.
6. Finish with a coat of clear varnish to give extra protection.

PRIMEVAL PAINTING : CHINA

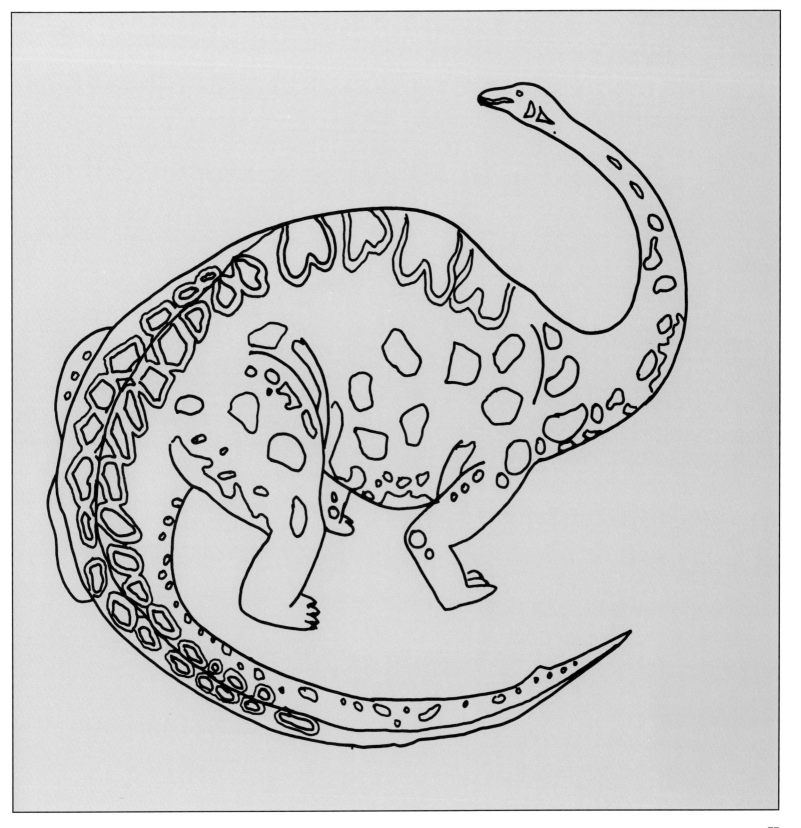

Memosaurus

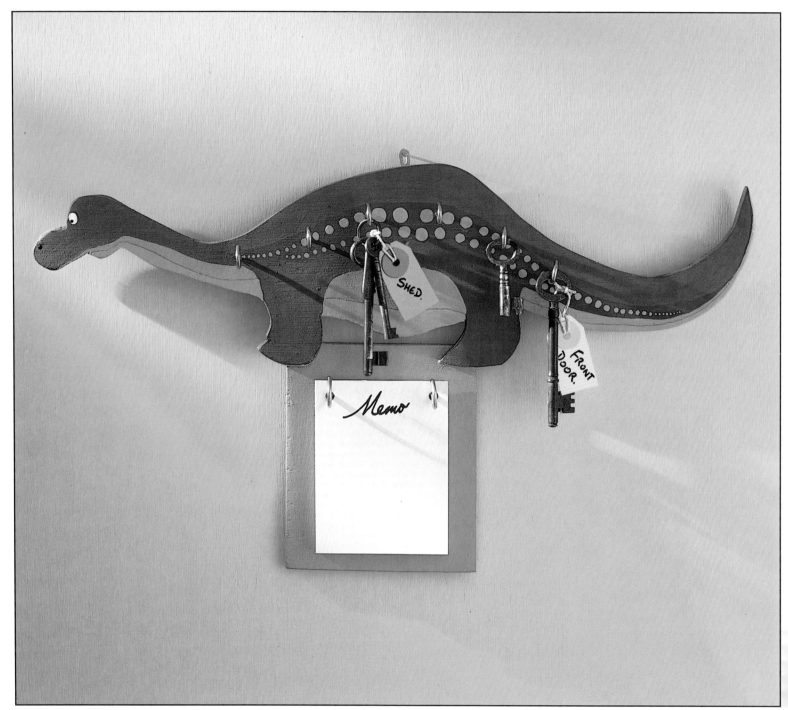

This gentle diplodocus acts as a very useful key holder, and holds a memo pad between its feet.

Materials

50 x 15cm of 12.5mm (1ft 7¾ x 6in of ½in) plywood
15 x 13cm of 5mm (6 x 5⅛in of ¼in) plywood
Enamel paints: dark green, light green, pink, white, black
Paint brushes
Memo pad
Fret-saw or electric skill saw
Hand-drill
1.5mm (1/16in) drill bit
8 brass hooks
Hanging eye
Panel pin
Hammer
Wood glue
Hole punch
Medium sandpaper

Method

1. Trace the dinosaur shape from the diagram, size up on graph paper, and transfer to the 50 x 15cm (1ft 7¾ x 6in) sheet of plywood. Cut out the shape with a fret-saw or skill saw.
2. With the saw, cut two notches into the reverse side of the dinosaur's feet. These will house the memo board.
3. Cut out and smooth with sandpaper the memo board (15 x 13cm – 6 x 5in sheet of plywood).
4. Sandpaper the dinosaur shape smooth.
5. Using the wood glue, glue the memo board to the reverse side of the feet, slotting it into the notches. Leave until the glue is dry.
6. Prime the dinosaur and memo board, and when dry, paint with enamel paints. Do the large areas of colour first, and leave to dry thoroughly before attempting the decorative spots, the eye and the nose. The latter is black, the eye is white with a black centre.
7. When the paint is dry and hard, lightly mark with a pencil the positions in which the hooks for the keys and memo pad are to be placed. Drill with a 1.5mm (1/16in) drill bit and screw in the hooks.
8. Hold the memo pad in position over its hooks and mark the positions. Make holes with a hole punch so the pad can be hung on the hooks.
9. Fix the hanging eye to the centre of the reverse side of the dinosaur's back with a panel pin.

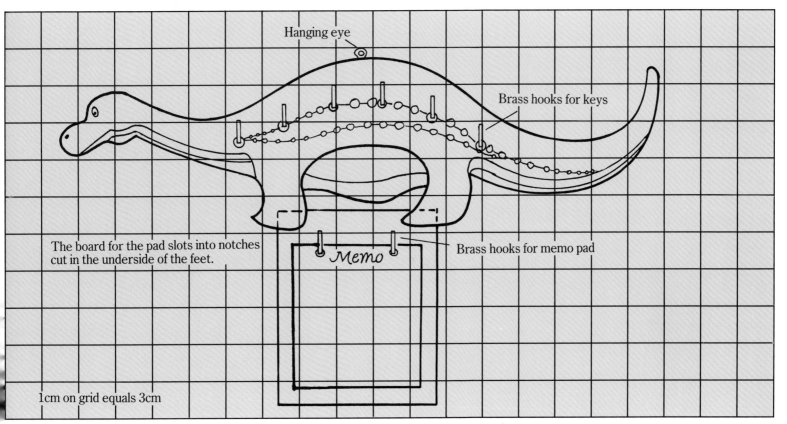

The board for the pad slots into notches cut in the underside of the feet.

1cm on grid equals 3cm

Stencil-osaurus

STENCIL-OSAURUS

Making your own stencils is easy and fun to do, and the children will be delighted to help. They can be used to decorate a variety of items – book covers, T-shirts, lampshades – or to make a frieze round a child's bedroom.

Materials

Stencil paper or board
Stencil brush
Acrylic or gouache paints
Scalpel or craft knife
Carbon paper
2H and 2B pencils
Fine waterproof felt-tip pen or rapidograph
Cutting board or other hard smooth surface
30cm (12in) ruler
Masking tape
Hole punch

Method

1. Decide on your design and scale it up or down if necessary. Although generally large designs look better on large areas, it can be fun sometimes to put a large design on a small area, such as a book jacket, and wrap it right round the front and back.
2. Tape the stencil paper or board onto the cutting board. Place the drawing of the dinosaur in the middle of the card, leaving 60mm (2½in) all round the design. This will give the stencil strength and stop it tearing.
3. To transfer the design to the stencil paper or board, use carbon paper or simply rub over the back of the design with a 2B pencil. Tape down the design and trace it carefully onto the stencil paper or board using a 2H pencil.
4. Cut out the design, using a fresh blade. If using a craft knife, you might find a round-handled one easier to use than a flat-handled one.
5. Holes for eyes can be made with a hole punch.

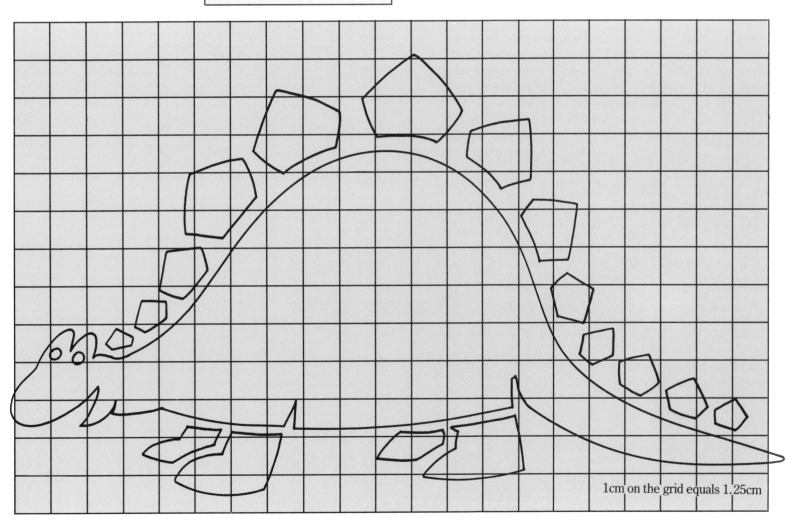

1cm on the grid equals 1.25cm

STENCIL-OSAURUS

6. Positioning the cut-out stencil. If you are just making a single stencil, you can position the stencil by eye. But if you want to repeat the pattern, for example, to make a frieze, you need to make a few preparations.

If your repeat pattern is to go on a roll of paper, tape the roll onto the work surface. Decide where on the roll the design is to go, and draw a light baseline with a 2B pencil, so you can rub it out when the stencilling is complete. If you are stencilling straight onto a wall, decide where the design is to go and draw a baseline in a similar way.

Then draw a line on the stencil across the bottom of the design. Line this up with the line on the paper roll or wall. Work out how many repeats will fit in the space you have, and mark the positions where the stencil will fall on the roll of paper or wall.

7. Stencilling. Mix the paint to a creamy consistency. Don't mix too much at a time. Before you start, do a test piece on a piece of paper similar to that on which you will be stencilling. Hold the brush as if it were a pencil and paint from the edges inwards, using soft stabbing strokes. Let your test piece dry before starting with the stencil. Make sure, when painting, that the paint isn't too runny or it will run underneath the stencil.

Position the stencil carefully and tape it down. When you've finished painting, let the paint dry before you remove the stencil card. Before putting the card down again, make sure it's clean, especially underneath. It's a slow business!

8. Painting more than one colour. Cover the parts of the design you don't want to paint with a piece of non-absorbent paper. Paint the area you wish to paint, then let it dry. Then use the paper to cover the painted parts while you paint the uncovered parts with a second colour. You can repeat the process as often as you like, but do make sure each part is dry before attempting the next.

9. If the stencil breaks, you can mend it with sticky tape, but be sure to cut off any bits that stick out and which might affect the design.

These stencils are actual size.

Alphabetical

You can use this wonderful dinosaur alphabet to make more complicated stencils, or simply to trace off a name or someone's initials.

 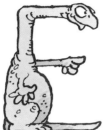    

   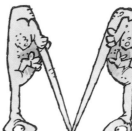  

ALPHABETICAL

BEDTIME DINOSAURS

# Hot-water Bird

This cosy hot-water bottle cover features an archaeopteryx, a bird-like dinosaur from 150 million years ago.

Materials

0.75m (¾yd) white polycotton
Two 38cm (15in) squares terylene wadding
0.75m of 1cm wide (¾yd of ½in wide) ribbon, cut in half
Fabric dye 'colourfun' paints or pens in red, blue, purple, yellow, green
'Puffy paint' in light green
Water-soluble fabric pen
Black cotton thread

Method

1. Scale up the design.
2. Cut four pieces of the fabric shape in the polycotton and two pieces in the wadding. Trim the wadding until it is 0.5cm (¼in) smaller than the fabric.
3. Using the water-soluble fabric pen, trace the design onto one piece of the fabric.
4. Using the fabric dye paints or pens, colour in the design. Apply the colours carefully. Let one colour dry completely before painting another colour next to it.
5. When the dye is dry, press the fabric with a hot iron to fix the colour.
6. Lastly, apply the light green 'puffy paint' to the edges of the leaves. When it is dry, turn the fabric over and iron it. The dye puffs up when heated.
7. Place a piece of wadding onto the back of the dyed fabric and tack it in place.
8. Using black cotton thread, hand- or machine-stitch round the archaeopteryx.
9. Pin another piece of fabric to the design, with right sides together.
10. Stitch through all the thicknesses 0.5cm (¼in) from the edge from (A) to (B) round the longer side, leaving the top open.
11. Turn right side out. These two pieces form the outer covering.
12. Slip the other piece of wadding inside the cover.
13. Put the two remaining pieces of polycotton right sides together and stitch round 0.5cm (¼in) from the edge from (A) to (B) as before to form the lining.
14. Slip the lining inside the cover – the wadding should now be sandwiched between the two layers of fabric on both sides.
15. Turn in the raw edges 0.5cm (¼in) and slip-stitch the lining to the outer bag.
16. Attach the ribbons to the top of the bag, one to the front and one to the back.

Hot-water Bird is shown on the previous page.

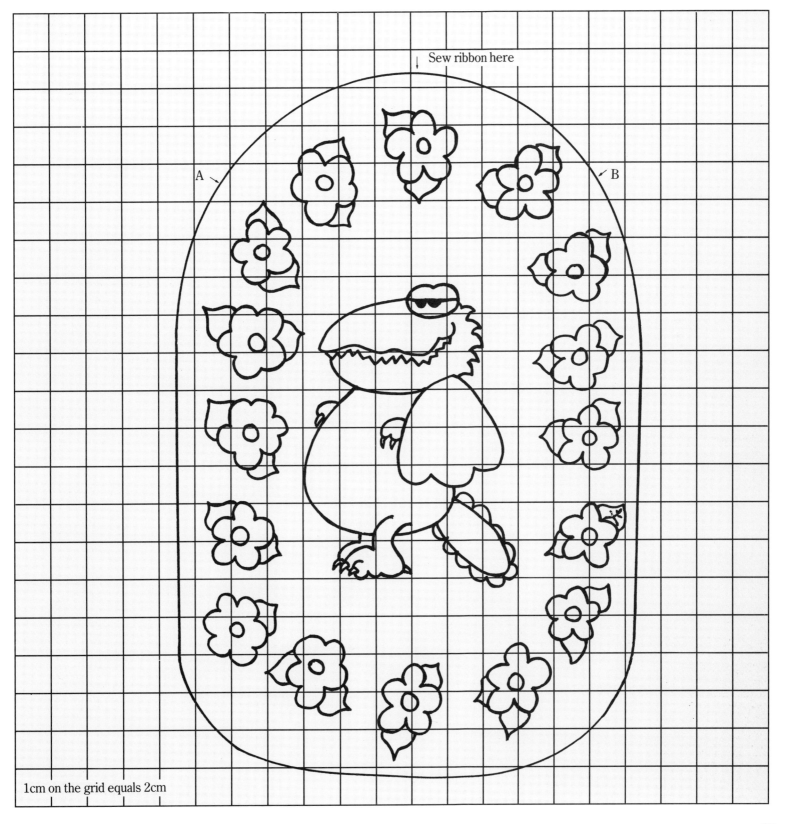

Sleeping Partner

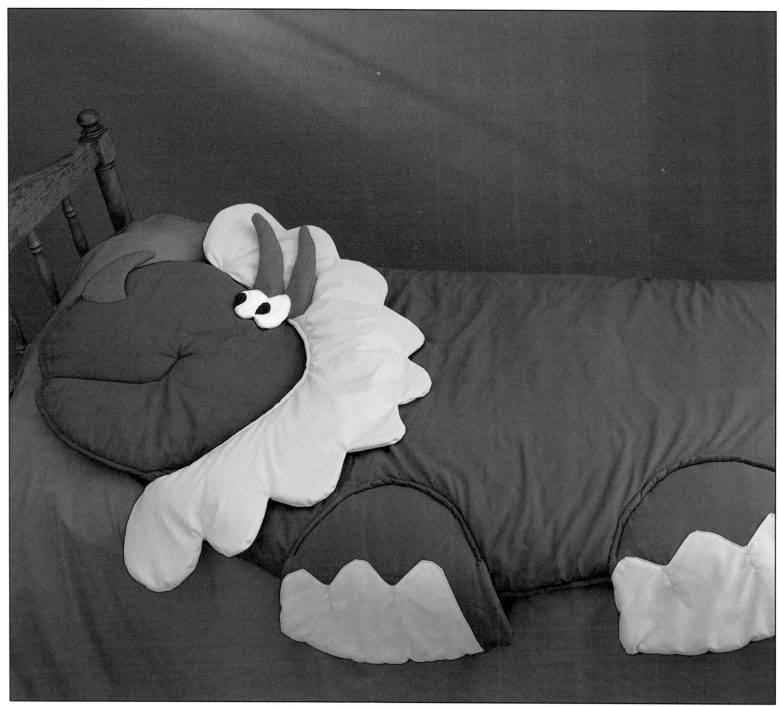

This jolly triceratops makes a delightful sleeping-bag for a child. Its head is the pillow, and the sleeper slides in under its spiky yellow collar.

Materials

8m of 125cm wide (8¾yd of 48in) purple cotton or polycotton
1m of 90cm wide (1yd of 36in) yellow cotton or polycotton
0.25m of 90cm wide (¼yd of 36in) pink cotton or polycotton
Scrap of white fabric for eyes
5m of 2.5cm wide (5½yd of 1in) purple bias binding
1m of 2.5cm wide (1yd of 1in) yellow bias binding
5m of 125cm wide (5½yd of 48in) thick terylene wadding
Tailors' chalk or air-soluble dressmaking pen
Fabric-dye pen or black washable paint

Method

1. Scale up all the pattern pieces.
2. Cut out two main body pieces (A), two upper body pieces (B) and four leg pieces (C) from purple fabric.
3. Cut out two collar pieces (D) and two toenail pieces (E) from yellow fabric.
4. Cut out two eye pieces (F) from white fabric.
5. Cut out four large horn pieces (G) and two small horn pieces (H) from pink fabric.
6. Cut out one upper body piece (B), one lower body piece (A), two leg pieces (C), one collar piece (D) and two head pieces (I) from the wadding.
7. Pin the wadding to the wrong side of one main body piece (A). Position and pin the two head pieces of wadding (I) to the head part of the main body to form the pillow. Pin the other main body piece on top, wrong side down. Machine stitch all round the main body, 1cm (½in) from the edge through all thicknesses.
8. Draw, with tailors' chalk or air-soluble pen, the outline of the head on the upper main body piece. Machine this line through all thicknesses. This forms the pillow.
9. Sandwich the wadding between the wrong sides of the upper body pieces. Pin, tack and machine stitch all round 1cm (½in) from the edge through all thicknesses.
10. Turn in the toenail edges as shown on the drawing 1cm (½in) to wrong side. Pin and tack the toenail to the right side of one leg piece as shown. Machine along the neatened edge, stitching down and back up again between the 'nails' as shown.
11. Place one plain leg piece and one leg piece with toenails right sides together. Pin the wadding on top. Machine stitch through all thicknesses between the large dots as shown on the pattern. Turn the leg to the right side and slip-stitch the seam closed.
12. Repeat with the other leg.
13. Place the legs on the upper body. Pin and tack in position then machine stitch through all thicknesses round the top of the legs, starting and stopping 6cm (2½in) from the edge of the body.
14. Place the collar pieces right sides together. Pin the wadding on top. Machine stitch through all thicknesses round the scalloped edge. Snip the corners and turn right side out.
15. Place the collar on top of the upper body at the neck edge and tack in place.
16. Pin the yellow bias binding on top of the collar with the binding edge along the edge of the collar and body. Machine stitch along the crease line of the binding ½cm (¼in) from its edge through all thicknesses. Turn the binding to the inside and slip-stitch in place.
17. Pin the upper body on top of the lower body, matching the edges. Trim if necessary. Pin the collar firmly out of the way.
18. Pin and tack the purple bias binding all round the upper body piece and around the head of the lower body. Machine stitch along the crease line of the binding ½cm (¼in) from its edge through all thicknesses. Turn the binding to the back of the body piece and slip-stitch in place.
19. Place the horn pieces (H) right sides together and pin the wadding on top. Machine stitch through all thicknesses almost all the way round, leaving a gap for turning.
20. Turn to the right side and slip-stitch the open seam.
21. Pin the horn in position on the head. Slip-stitch along its bottom edge.
22. Make two more horns from pieces (G) in the same way. Position these onto the collar and sew in place.
23. Pin the eye pieces right sides together and pin the wadding on top. Machine through all thicknesses almost all the way round, leaving a gap for turning.
24. Turn to the right side and slip-stitch the open seam.
25. With an air-soluble marker pen draw the dividing line between the eyes and machine stitch this line.
26. With a fabric-dye pen, colour in the pupils of the eyes. Position the eyes on top of the horns and collar and stitch in place.
27. With an air-soluble marker pen draw the line of the mouth and machine stitch this line.

SLEEPING PARTNER

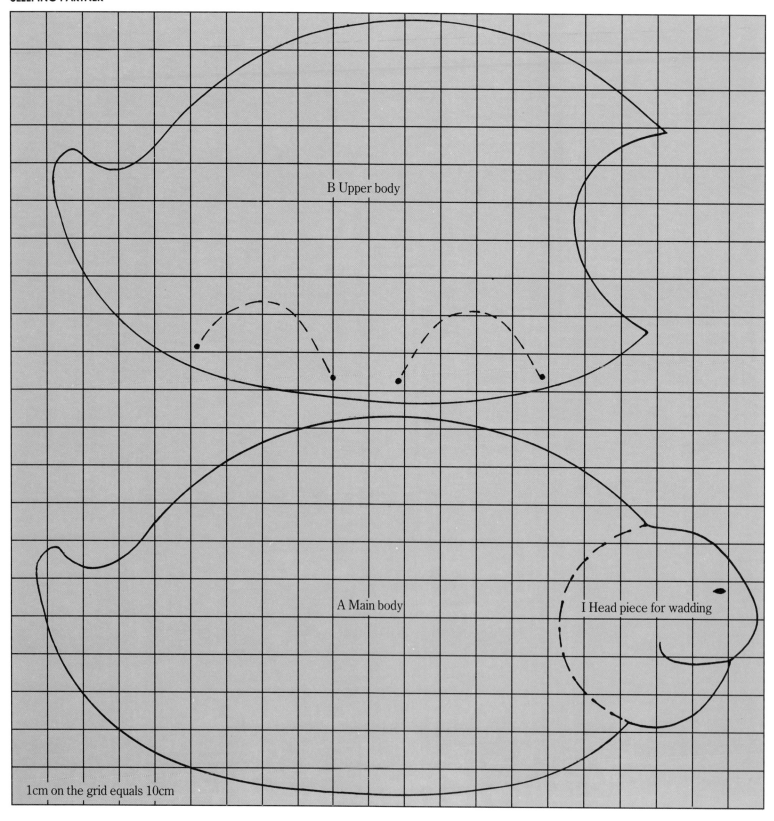

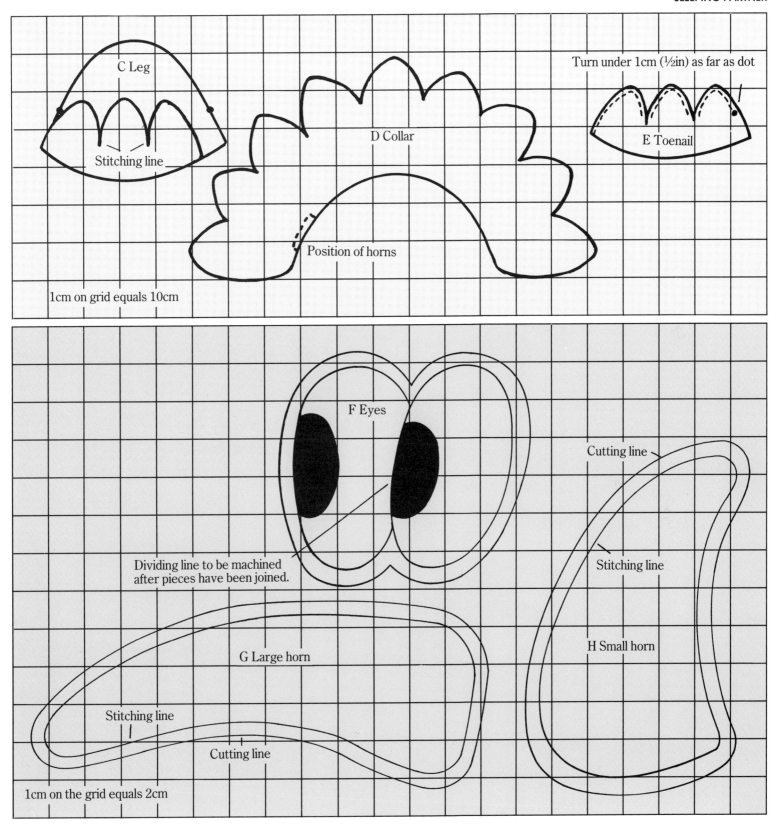

Bronto-snaurus

BRONTO-SNAURUS

This sleepy brontosaurus makes a bedspread which will be a delightful addition to a child's bedroom.

Materials

Approx. 240 x 170cm (95 x 69in) polycotton sheeting fabric for single bedspread Plus 60 x 110cm (24 x 43in) of the same for a matching panel

0.75m of 120cm (⅚yd x 48in) interfacing

60 x 110cm of 70g (24 x 43in of 2½oz) polyester wadding

Approx. 71 x 43cm (28 x 17in) pink polycotton for appliqué

61 x 38cm (24 x 15in) purple polycotton for appliqué

24 x 30cm (10 x 12in) yellow polycotton for appliqué

Scrap of main colour for eye

Tailors' chalk

Method

1. Scale up the pattern.
2. Cut out the large piece of fabric for the bedspread and the smaller piece for the matching panel. (If you want a contrasting panel you could use a single sheet for the bedspread).
3. Iron the materials to be used for appliqué onto the interfacing, then draw the shapes carefully onto the interfacing. If you are using a paper template you will find it easier to tape the paper to the interfacing at intervals to prevent movement.
4. Cut the pieces for the appliqué, taking care not to cut the 'stitching lines' or 'overlapping lines'.
5. Pin the wadding to the panel piece and zig-zag round the edges to secure. Trim the wadding if necessary.
6. Find the centre of the panel by folding it in half in both directions and mark the centre with a pin.
7. Mark with a pin the centre of the purple underbody of the dinosaur (the centre is marked on the pattern), and place the underbody on the panel, matching the centres. Pin the shape in place, taking care to keep it smooth and straight.
8. Stitch the shape in place by first top-stitching from the top of its head, down its back to the tip of its tail, then by satin-stitching around its front, using stitch width 3.
9. Place the pink body on top of the underbody, using the overlap lines as a guide. Ensure that the pieces meet at the tip of the tail and where the far hindleg appears to go behind the underbody (see diagram).
10. Mark with chalk on the pink shape the 'stitching lines' shown on the pattern.
11. Pin the pink shape in place, taking care to keep it smooth and straight. Satin-stitch around it, and along the 'stitching lines' you have chalked on.
12. Pin the nightcap, candle, candle-holder, candle-flame and eye in place, with the candle-holder slightly overlapping the candle. Satin-stitch round all the shapes, stitching the candle first and then the candle-holder over it.
13. Turn in the edges of the panel, making sure you hide all the zig-zag stitch, and machine-stitch all round approximately 1cm (½in) from the edge. Mitre the corners to neaten them.
14. To make the bedspread, take the large piece of material and machine-stitch the hems all round, making the top one deeper.
15. Find the centre of the bedspread by folding it in half in both directions. Mark the fold lines with chalk.
16. Find the centre of the panel in the same way, and mark the fold lines.
17. Match the fold lines of the panel and bedspread, putting the reverse of the panel on the right side of the bedspread. Pin at the marked points first (see diagram), ensuring the material is completely flat under the panel.
18. Tack the panel in place. Machine top-stitch between the edges of the panel and its hem, to give a piped effect.

BRONTO-SNAURUS

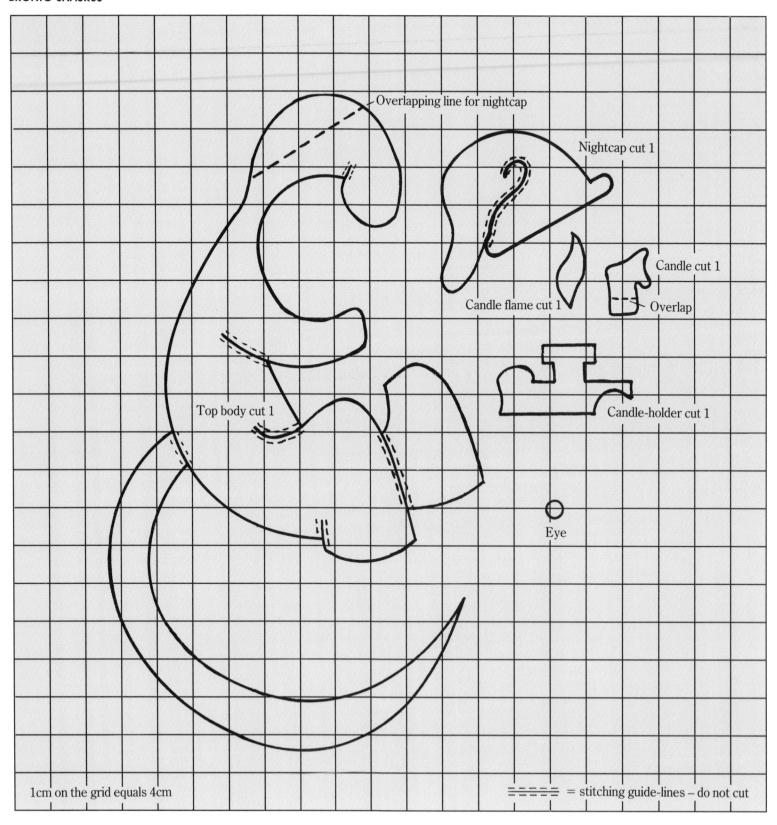

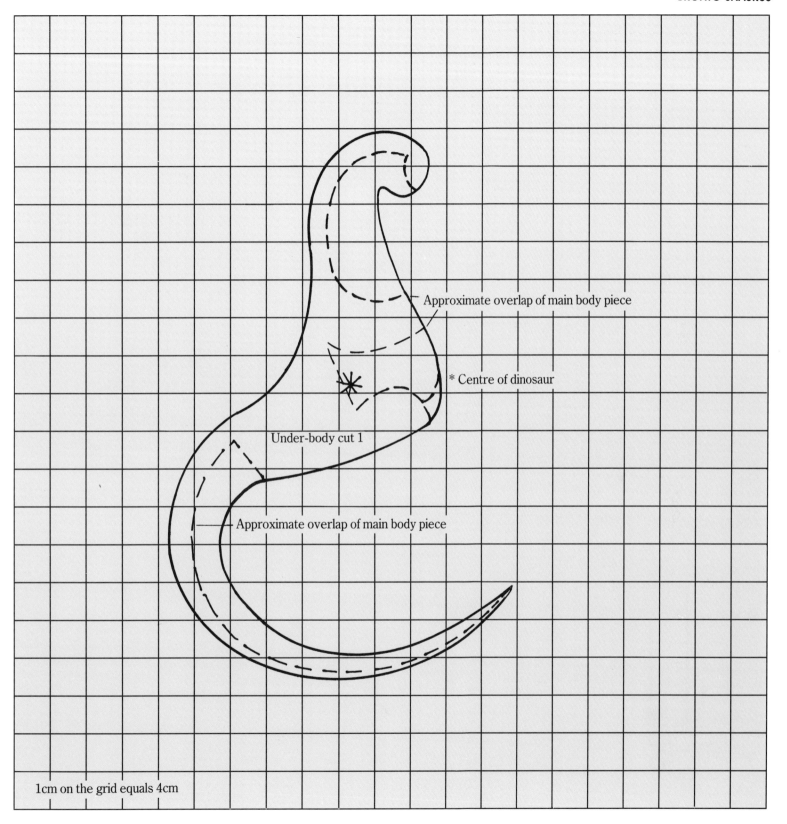

BRONTO-SNAURUS

The top body and underbody shapes must meet at these points.

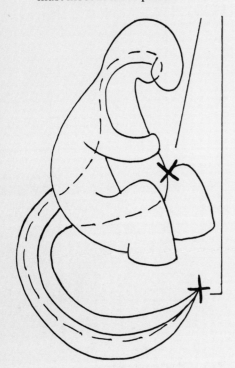

How to assemble the pieces.
The dotted lines show the overlap.

Top-stitching

Satin-stitching

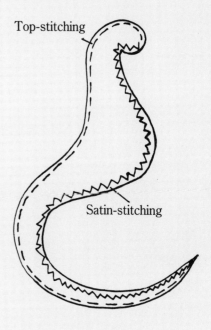

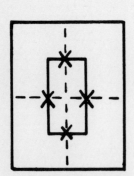

Pin the panel on the bedspread at the marked points (X) first.

PARTY-TIME DINOSAURS

Bertie Baryonynx

Open this pop-up card and you are confronted with the mighty jaws of Bertie, the bold baryonynx.

Materials

An A4 sheet of card, folded in half to make A5*
An A4 sheet of pink paper
An A4 sheet of green paper
A small piece of white paper
Scissors
Pencil
Black felt pen
Paper glue
*or smaller, if desired, in which case the dinosaur should be made smaller.

Bertie Baronynx is shown on the previous page.

Method

1. Scale up the pattern.
2. Put a sheet of pink paper over a sheet of green paper, and cut out the main shape, (A), from these two. The paper should not be too thick or it will be difficult to fold.
3. Cut out (B) from the green paper only. Draw the eye shapes on this with the black felt pen.
4. Cut out the horn shape, (C), from white paper.
5. Glue together the nose of the dinosaur (i) to 1, and the chin, (ii) to 2. The green paper is the outer skin; the pink inside the mouth.
6. Then glue (iii) to 3 and (iv) to 4.
7. From the underneath, push the eyes into place through slots A and B.
8. Fold over the horn piece, (C), and glue it in shape. Push it through slot C and glue it in place by folding back the base onto the dinosaur's nose.
9. Find the centre of the folded white card and draw on it lightly in pencil the diamond shape which forms the back of the head of the dinosaur. Glue the head onto the card and leave it to dry.
10. When quite dry, fold the card up and open it again. Bertie will leap into life.

Note: if desired, the outside of the card could be stencilled with one of the shapes on pages 81 and 82, or a message written on it in the dinosaur alphabet on pages 83 and 84.

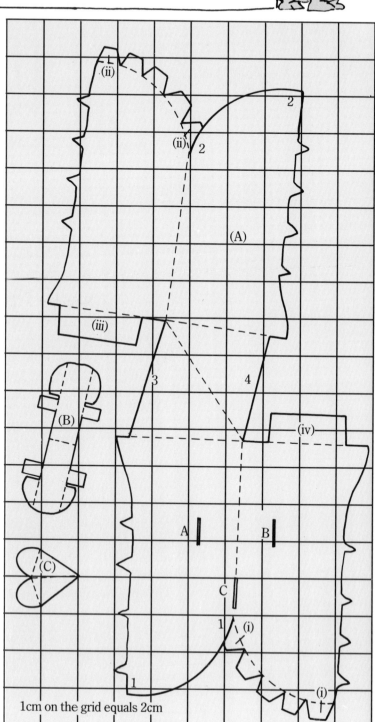

1cm on the grid equals 2cm

BERTIE BARYONYX

How to assemble the pop-up card

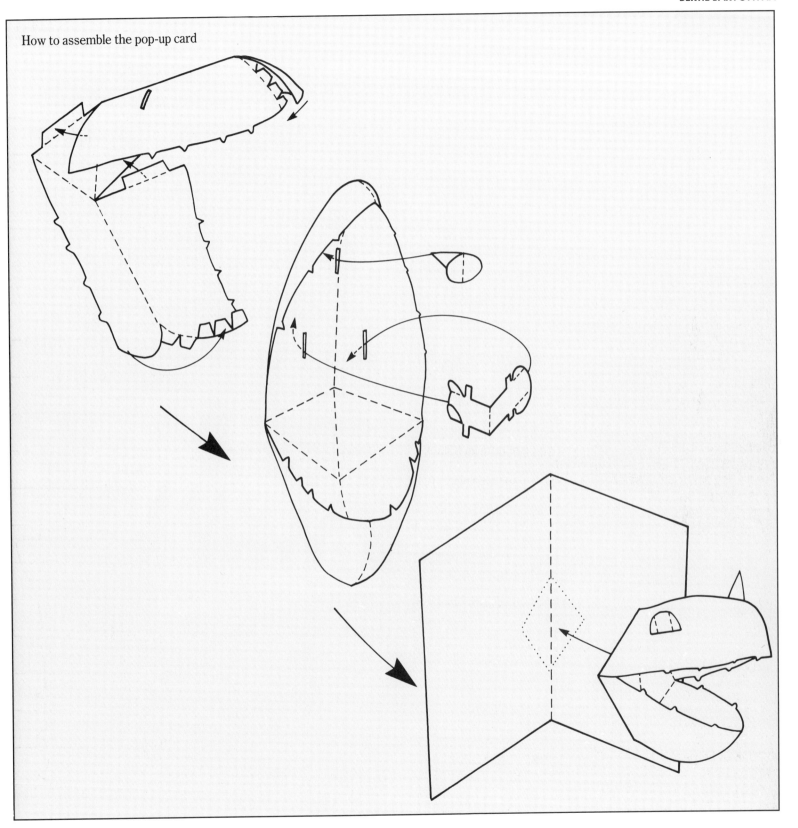

Monster Mask

MONSTER MASK

This toothy tyrannosaurus mask fits most sizes, and is ideal for a fancy dress party. The wearer looks out through its mouth.

Materials

1 A2 and 1 A3 sheets of stiff orange paper (16½ x 24in and 16½ x 12in)
1 A2 (16½ x 24in) sheet of stiff pink paper
2 A3 (16½ x 12in) sheets of stiff white paper
Small piece of stiff black paper
Approximately 1m (1yd) 1cm (½in) wide pink or orange ribbon
Pencil
Scissors
Paper glue
Hole punch

Method

1. Size up the pattern.
2. Cut out the double piece (A) from orange paper.
3. Cut out piece (B) from pink paper.
4. Cut out pieces (C) and (D) from orange paper.
5. Cut out a small black circle to fit the inner circle of (C) and glue it in place to form the centre of the eye.
6. Cut out a circle of white to fit the centre ring of (C) to form the white of the eye and glue it in place.
7. On (A), glue (i) to 1 and (ii) to 2.
8. Glue together the nose of the dinosaur, (iii) to 3.
9. Glue together the back of the head (iv) to 4.
10. Gently fold the right eye so that the pupil sticks out, then glue (v) to 5.
11. Fold in the tabs (vi) round the eye socket on (A) inwards, then insert the eye into the socket, gluing (vi) to 6.
12. Assemble and position the left eye in the same way.
13. Glue the part (B) into position, sticking (vii) to 7.
14. Glue the band (D) into position, sticking (viii) to 8.
15. Cut out the whole lower jaw shape (E) in pink and orange paper by flipping the pattern over from the centre line.
16. Cut out the whole of the lower teeth shapes in white paper in the same way.
17. Glue the pink jaw shape behind the orange, and the white teeth shapes over the orange teeth shapes. Fold the teeth back and then forward along the dotted lines.
18. Cut out the whole upper jaw shape (F) in white paper by flipping the pattern over from the centre line.
19. Glue the upper teeth onto the lower part of the upper jaw by means of the tabs as shown in the assembly diagram.
20. Glue the lower jaw onto the 'cheeks' of the dinosaur head as shown in the assembly diagram.
21. Punch a hole at the centre of (D).
22. Loop a piece of ribbon through (D), leaving the ends free to tie under the chin and secure the mask when it is being worn.

MONSTER MASK

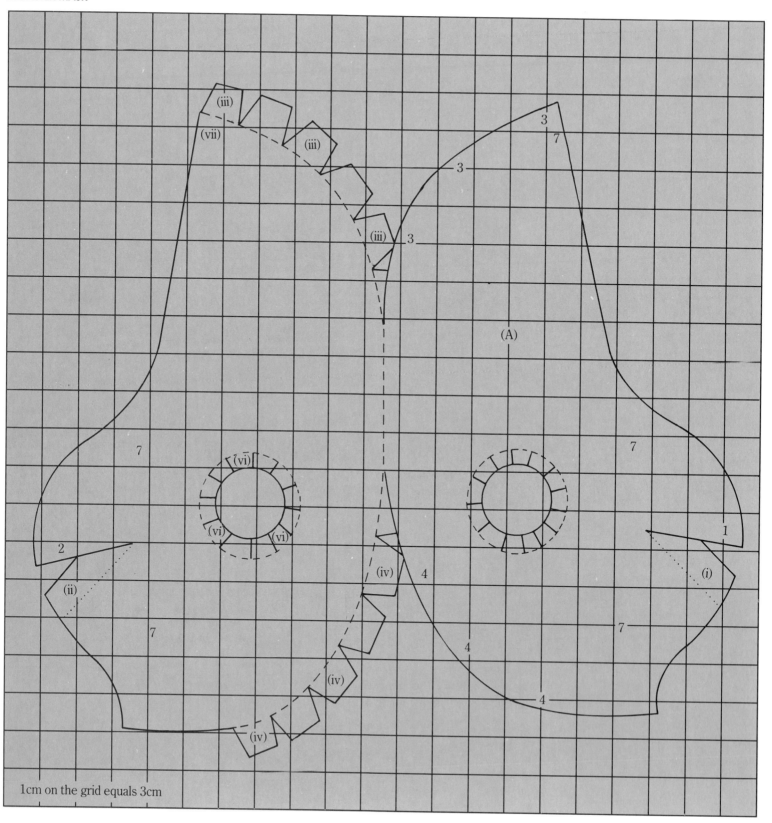

1cm on the grid equals 3cm

102

MONSTER MASK

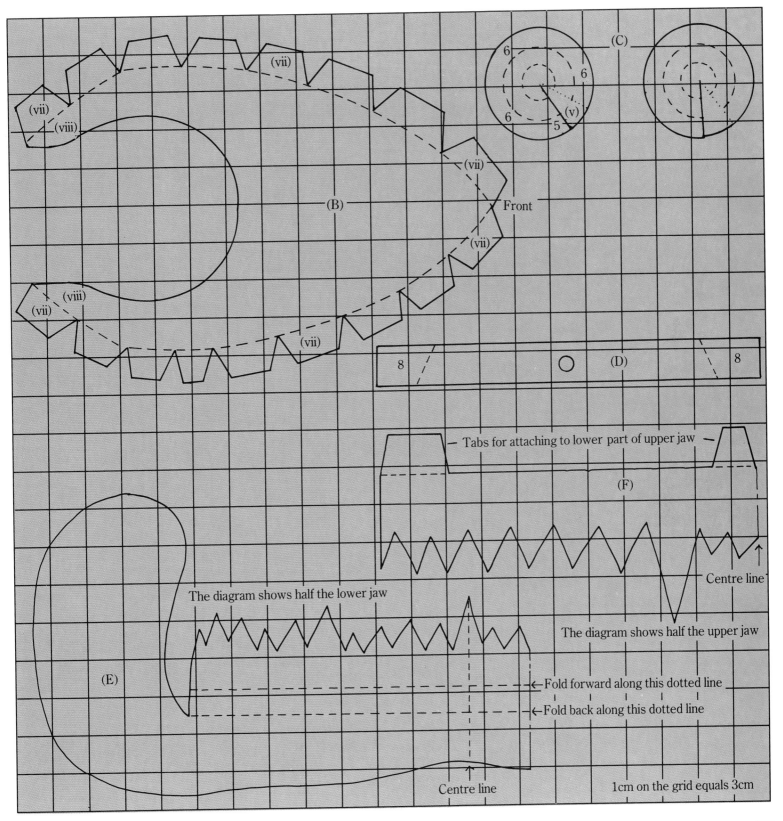

MONSTER MASK

How to assemble the mask

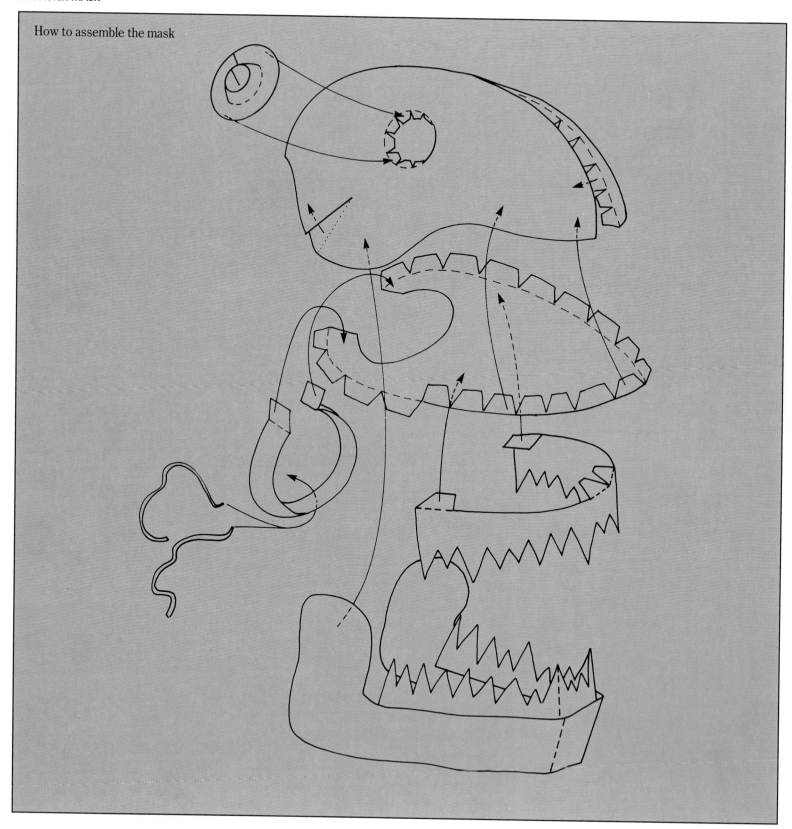

Party-time Stegosaurus

PARTY-TIME STEGOSAURUS

This spiky stegosaurus is made from a simple madeira cake recipe, making it a firm favourite for children's parties.

Materials

19 x 27cm (7½ x 10½in) rectangular madeira cake
300g (12oz) buttercream
1kg (2lb) sugarpaste
Blue, orange and green food colourings
Icing sugar for dusting

38cm (15in) square cake board
Curved crimpers or teaspoon
Sweets for decoration
10 blanched almonds
1 cherry
Birthday candles, if desired

To Make the Cake

Ingredients
250g (8oz) (1 cup) soft butter
250g (8oz) (2 cups) caster (superfine) sugar
4 eggs, size 2
250g (8oz) (2 cups) self-raising flour
125g (4 oz) (1 cup) plain flour
Grated rind 1 lemon
1 tbsp lemon juice
1-2 tbsp milk

Method

1. Preheat the oven to 160°C, 325°F, gas mark 3.
2. Grease and base line a rectangular tin measuring 19 x 27 x 5cm (7½ x 10½ x 2in).
3. Warm the bowl and beater. Place the whole eggs in a bowl of warm water. Sift the flours together.
4. Cream the butter and sugar until light and fluffy. Break the eggs and beat them in a little at a time, adding 1 dessertspoonful of flour with each egg.
5. Using a metal spoon, gently fold in the remaining flour, together with the lemon rind and juice. If necessary, add a little milk to make the mixture have a dropping consistency.
6. Turn into the prepared tin and bake for about one hour or until golden brown and firm to the touch.
7. Leave to cool on a wire rack for ten minutes before turning out and removing the lining paper.

To Make the Buttercream

Ingredients
150g (6oz) (¾ cup) unsalted butter
150g (6oz) (1½ cups) icing (confectioners') sugar
2 tbsp lemon or orange juice

Method

1. Beat the softened butter until light.
2. Gradually add the sifted icing sugar, beating well after each addition.
3. Beat in the flavouring until well blended.

To Assemble the Dinosaur Cake

1. Make the templates (A), (B), (C), (D), (E), (F) and (G) and place on top of the cake. Using a sharp knife, cut round the templates.
2. Slice horizontally through pieces (B), (C), (D) and (E), to make two sides, two thighs and four feet.
3. Place the head and body piece, (A), diagonally across the board and spread it with a thin layer of buttercream.
4. Position the sides, (B), on each side as shown in the diagram, and spread with buttercream.
5. Attach the thighs, (C), to the side body pieces (B), as shown.
6. Slice horizontally through the tail pieces (F) and (G). Using one layer of each section, attach to the body with buttercream. Trim to a rounded shape with a pointed end, and curve the tail forward on the board.
7. Using a sharp knife, shave off the top edges of the body to give a rounded appearance. Smooth buttercream over the entire surface.

PARTY-TIME STEGOSAURUS

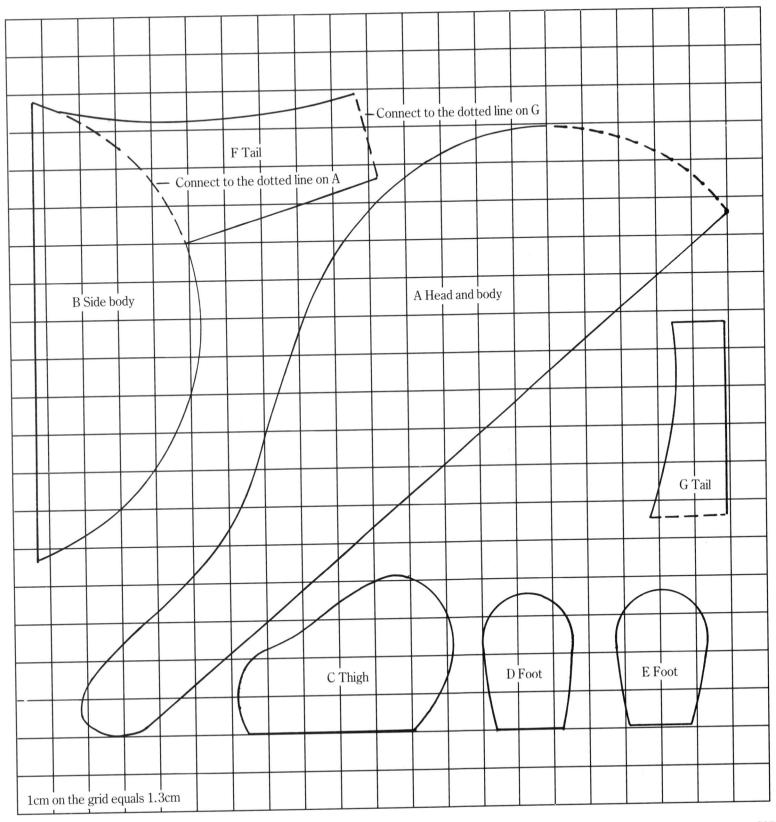

107

PARTY-TIME STEGOSAURUS

8. Roll out 750g (1½lb) sugarpaste to a strip measuring 53 x 27cm (21 x 11in), tapering each end. Mark the sugarpaste with crimpers or the tip of a teaspoon to give a scaly effect.

9. Lay the sugarpaste strip over the cake and smooth round the curves of the head and tail. Trim, and mould into the base.

10. Spread the feet pieces, (two of (D) and two of (E)) with buttercream and cover with sugarpaste. Trim, and position on either side of the body, as shown in the diagram.

11. Dilute the blue food colouring in 1 dessertspoonful of water and brush it over the dinosaur.

12. Cut the almonds lengthways and position as claws on the feet. Stick four in as spikes in the end of the tail.

13. Colour 100g (4oz) sugarpaste orange. Roll out to a strip about 56 x 8cm (22 x 3in) tapering at each end.

14. Moisten a narrow strip along the top of the body with water and lay the sugarpaste on it. Use a small, sharp knife to cut out triangular shapes.

15. With a pointed knife, make slits along the back of the body and position the sweets in the sugarpaste to represent the spines. We used liquorice allsorts and jelly beans.

16. Cut two slivers of cherry and position them each side for eyes.

17. Colour the remaining white sugarpaste green. Press small pieces through a metal sieve and cut away the strands with a knife.

18. Soften the rest of the green sugarpaste to a spreading consistency with a little water, and use it to cover the board. Position the strands of sugarpaste to resemble grass. Add small coloured sugarpaste flowers if desired.

19. Arrange the candles in the curve of the tail.

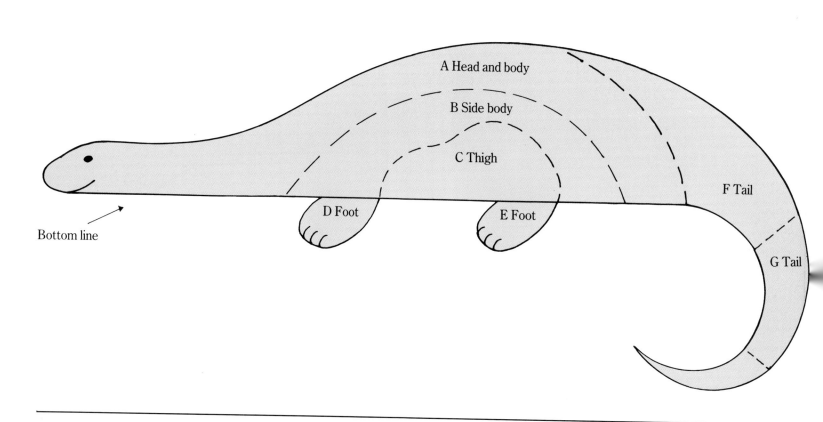

Eggs in the Nest

EGGS IN THE NEST

These charming eggs are hollow, and can be used to conceal little marzipan baby dinosaurs, to delight young dinosaur lovers. All the eggs are edible.

Materials

- 1 9cm (3½in) plain Easter egg mould
- 150g (6oz) (1⅓ cup) caster (superfine) sugar
- 1kg (2lb) sugarpaste
- 250g (8oz) (2 cups) brown sugar
- 50g (2oz) white marzipan
- Flaked almonds
- 1 cherry
- Red, blue, green and mauve food colourings
- Icing sugar for dusting

- New toothbrush
- Small piece of sponge
- 30cm (12in) round cake board
- Red ribbon

Method

Sparkling egg
1. Mix about 1 teaspoonful of water with a little mauve food colouring and add 150g (6oz) caster sugar. The sugar should be the consistency of damp sand.
2. Pack the sugar into the egg mould and level the top surface with a knife.
3. Unmould the egg immediately on a flat surface and leave to dry for a few hours.
4. Once the outside surface is firm, carefully return the egg to the mould and scoop out the soft sugar, leaving a 6mm (¼in) shell. Remove from the mould and leave to harden.
5. Mould the second half in the same way.
6. Join the two halves together with softened sugar and fill in any gaps in the join.

Coloured sugarpaste eggs
1. Allow 100g (4oz) sugarpaste for each egg. Colour it green or as desired.
2. Roll out the sugarpaste to 1cm (¼in) thick. Dust the inside of the mould with icing sugar. Line the mould with sugarpaste and remove the surplus from the top edge with a sharp knife.
3. Gently move the sugarpaste to ensure it is not sticking to the mould, then leave it to dry in the mould for a few hours, then remove carefully from the mould and leave to dry completely.
4. Stick the two halves of the egg together with sugarpaste, softened to a spreading consistency with water.

Blue speckled egg
1. Colour 100g (4oz) sugarpaste pale blue and make an egg as before and leave to dry.
2. Dip a new toothbrush into darker blue food colouring, and using a fingertip to pull the bristles back, spray colour onto the egg's surface.

Broken egg
1. Line the mould, and while the sugarpaste is still soft, use a sharp knife to cut away a jagged section of the egg. Ensure that the hole is large enough for the baby dinosaur to break through.

Baby dinosaur
1. Roll the marzipan into a sausage shape and slice two pieces for the front legs with a knife. Using fingers, model the head and legs.
2. Cut the claw shapes with scissors and insert tiny pieces of almonds to make them.
3. Slit the mouth with a knife.
4. Colour a tiny piece of marzipan red. Flatten it out, and cut out a forked tongue. Insert the tongue into the mouth.
5. Use tiny pieces of cherry to make eyes.
6. Leave to dry for twelve hours, then use a small sponge to stipple green food colouring lightly over the body
7. Carefully insert the baby into the broken egg.

Nest
1. Roll 500g (1lb) sugarpaste into a long sausage.
2. Dampen the cake board with water, and make the sugarpaste into a circle on it.
3. Brush the circle with water, and sprinkle the board and sugarpaste with brown sugar.
4. Position the eggs within the circle.
5. Glue the ribbon round the edge of the board.

Acknowledgements

The publishers would like to thank the following people for
the help they have given in the preparation of this book:
Michèle Brown, Cortina Butler, Jenny Noll.